MW00667298

SOURCE

The Prentice Hall
ENGINEERING SOURCE

Introduction to Mechanical Engineering

Robert Rizza

*Department of Mechanical Engineering
and Applied Mechanics
North Dakota State University - Fargo, ND*

Prentice Hall
Upper Saddle River, NJ 07458

Library of Congress Cataloging-in-Publication Data

Rizza, Robert, 1965–
    Introduction to mechanical engineering / by Robert Rizza.
       p. cm. — (ESource—the Prentice Hall engineering source)
    Includes bibliographical references and index.
    ISBN 0–13–019640–1
    1. Mechanical engineering. I. Title. II. Series.

TJ145.R59 2000
621—dc21

00.059829

Vice-president of editorial development, ECS: **MARCIA HORTON**
Acquisitions editor: **ERIC SVENDSEN**
Associate editor: **JOE RUSSO**
Vice-president of production and manufacturing: **DAVID W. RICCARDI**
Executive managing editor: **VINCE O' BRIEN**
Managing editor: **DAVID A. GEORGE**
Editorial/production supervisor: **AUDRI ANNA BAZLEN**
Cover director: **JAYNE CONTE**
Art editor: **ADAM VELTHAUS**
Manufacturing buyer: **PAT BROWN**
Editorial assistant: **KRISTEN BLANCO**
Market manager: **DANNY HOYT**

© 2001 by Prentice-Hall, Inc.
Upper Saddle River, New Jersey 07458

10 9 8 7 6 5 4 3 2 1

ISBN 0-13-019640-1

Prentice-Hall International (UK) Limited, *London*
Prentice-Hall of Australia Pty. Limited, *Sydney*
Prentice-Hall Canada, Inc., Toronto
Prentice-Hall Hispanoamericana, S.A., *Mexico*
Prentice-Hall of India Private Limited, *New Delhi*
Prentice-Hall of Japan, Inc., *Tokyo*
Pearson Education (Singapore) Pte. Ltd., *Singapore*
Editoria Prentice-Hall do Brasil, Ltda., *Rio de Janeiro*

About ESource

ESource—The Prentice Hall Engineering Source
—www.prenhall.com/esource

ESource—The Prentice Hall Engineering Source gives professors the power to harness the full potential of their text and their first-year engineering course. More than just a collection of books, ESource is a unique publishing system revolving around the ESource website—www.prenhall.com/esource. ESource enables you to put your stamp on your book just as you do your course. It lets you:

Control   You choose exactly what chapter or sections are in your book and in what order they appear. Of course, you can choose the entire book if you'd like and stay with the authors' original order.

Optimize   Get the most from your book and your course. ESource lets you produce the optimal text for your students needs.

Customize   You can add your own material anywhere in your text's presentation, and your final product will arrive at your bookstore as a professionally formatted text.

ESource ACCESS

Starting in the fall of 2000, professors who choose to bundle two or more texts from the ESource series for their class, or use an ESource custom book will be providing their students with complete access to the library of ESource content. All bundles and custom books will come with a student password that gives web ESource ACCESS to all information on the site. This passcode is free and is valid for one year after initial log-on. We've designed ESource ACCESS to provides students a flexible, searchable, on-line resource.

ESource Content

All the content in ESource was written by educators specifically for freshman/first-year students. Authors tried to strike a balanced level of presentation, an approach that was neither formulaic nor trivial, and one that did not focus too heavily on advanced topics that most introductory students do not encounter until later classes. Because many professors do not have extensive time to cover these topics in the classroom, authors prepared each text with the idea that many students would use it for self-instruction and independent study. Students should be able to use this content to learn the software tool or subject on their own.

While authors had the freedom to write texts in a style appropriate to their particular subject, all followed certain guidelines created to promote a consistency that makes students comfortable. Namely, every chapter opens with a clear set of **Objectives**, includes **Practice Boxes** throughout the chapter, and ends with a number of **Problems**, and a list of **Key Terms**. **Applications Boxes** are spread throughout the book with the intent of giving students a real-world perspective of engineering. **Success Boxes** provide the student with advice about college study skills, and help students avoid the common pitfalls of first-year students. In addition, this series contains an entire book titled ***Engineering Success*** by Peter Schiavone of the University of Alberta intended to expose students quickly to what it takes to be an engineering student.

Creating Your Book

Using ESource is simple. You preview the content either on-line or through examination copies of the books you can request on-line, from your PH sales rep, or by calling 1-800-526-0485. Create an on-line outline of the content you want, in the order you want, using ESource's simple interface. Either type or cut and paste your own material and insert it into the text flow. You can preview the overall organization of the text you've created at anytime (please note, since this preview is immediate, it comes unformatted.), then press another button and receive an order number for your own custom book. If you are not ready to order, do nothing—ESource will save your work. You can come back at any time and change, re-arrange, or add more material to your creation. You are in control. Once you're finished and you have an ISBN, give it to your bookstore and your book will arrive on their shelves six weeks after they order. Your custom desk copies with their instructor supplements will arrive at your address at the same time.

To learn more about this new system for creating the perfect textbook, go to www.prenhall.com/esource. You can either go through the on-line walkthrough of how to create a book, or experiment yourself.

Supplements

Adopters of ESource receive an instructor's CD that contains professor and student code from the books in the series, as well as other instruction aides provided by authors. The website also holds approximately **350 Powerpoint transparencies** created by Jack Leifer of Univ. of Kentucky—Paducah available to download. Professors can either follow these transparencies as pre-prepared lectures or use them as the basis for their own custom presentations.

Titles in the ESource Series

Introduction to UNIX
0-13-095135-8
David I. Schwartz

Introduction to AutoCAD 2000
0-13-016732-0
Mark Dix and Paul Riley

Introduction to Maple
0-13-095133-1
David I. Schwartz

Introduction to Word
0-13-254764-3
David C. Kuncicky

Introduction to Excel, 2/e
0-13-016881-5
David C. Kuncicky

Introduction to Mathcad
0-13-937493-0
Ronald W. Larsen

Introduction to AutoCAD, R. 14
0-13-011001-9
Mark Dix and Paul Riley

Introduction to the Internet, 3/e
0-13-031355-6
Scott D. James

Design Concepts for Engineers
0-13-081369-9
Mark N. Horenstein

Engineering Design—A Day in the Life of Four Engineers
0-13-085089-6
Mark N. Horenstein

Engineering Ethics
0-13-784224-4
Charles B. Fleddermann

Engineering Success
0-13-080859-8
Peter Schiavone

Mathematics Review
0-13-011501-0
Peter Schiavone

Introduction to C
0-13-011854-0
Delores Etter

Introduction to C++
0-13-011855-9
Delores Etter

Introduction to MATLAB
0-13-013149-0
Delores Etter with David C. Kuncicky

Titles in the ESource Series

About the Authors

No project could ever come to pass without a group of authors who have the vision and the courage to turn a stack of blank paper into a book. The authors in this series worked diligently to produce their books, provide the building blocks of the series.

Stephen J. Chapman received a BS in Electrical Engineering from Louisiana State University (1975), an MSE in Electrical Engineering from the University of Central Florida (1979), and pursued further graduate studies at Rice University. Mr. Chapman is currently Manager of Technical Systems for British Aerospace Australia, in Melbourne, Australia. In this position, he provides technical direction and design authority for the work of younger engineers within the company. He is also continuing to teach at local universities on a part-time basis.

Mr. Chapman is a Senior Member of the Institute of Electrical and Electronics Engineers (and several of its component societies). He is also a member of the Association for Computing Machinery and the Institution of Engineers (Australia).

Mark Dix began working with AutoCAD in 1985 as a programmer for CAD Support Associates, Inc. He helped design a system for creating estimates and bills of material directly from AutoCAD drawing databases for use in the automated conveyor industry. This system became the basis for systems still widely in use today. In 1986 he began collaborating with Paul Riley to create AutoCAD training materials, combining Riley's background in industrial design and training with Dix's background in writing, curriculum development, and programming. Dix and Riley have created tutorial and teaching methods for every AutoCAD release since Version 2.5. Mr. Dix has a Master of Education from the University of Massachusetts. He is currently the Director of Dearborn Academy High School in Arlington, Massachusetts.

Delores M. Etter is a Professor of Electrical and Computer Engineering at the University of Colorado. Dr. Etter was a faculty member at the University of New Mexico and also a Visiting Professor at Stanford University. Dr. Etter was responsible for

the Freshman Engineering Program at the University of New Mexico and is active in the Integrated Teaching Laboratory at the University of Colorado. She was elected a Fellow of the Institute of Electrical and Electronics Engineers for her contributions to education and for her technical leadership in digital signal processing. In addition to writing best-selling textbooks for engineering computing, Dr. Etter has also published research in the area of adaptive signal processing.

Charles B. Fleddermann is a professor in the Department of Electrical and Computer Engineering at the University of New Mexico in Albuquerque, New Mexico. All of his degrees are in electrical engineering: his Bachelor's degree from the University of Notre Dame, and the Master's and Ph.D. from the University of Illinois at Urbana-Champaign. Prof. Fleddermann developed an engineering ethics course for his department in response to the ABET requirement to incorporate ethics topics into the undergraduate engineering curriculum. *Engineering Ethics* was written as a vehicle for presenting ethical theory, analysis, and problem solving to engineering undergraduates in a concise and readily accessible way.

Acknowledgments: I would like to thank Profs. Charles Harris and Michael Rabins of Texas A & M University whose NSF sponsored workshops on engineering ethics got me started thinking in this field. Special thanks to my wife Liz, who proofread the manuscript for this book, provided many useful suggestions, and who helped me learn how to teach "soft" topics to engineers.

Kirk D. Hagen is a professor at Weber State University in Ogden, Utah. He has taught introductory-level engineering courses and upper-division thermal science courses at WSU since 1993. He received his B.S. degree in physics from Weber State College and his M.S. degree in mechanical engineering from Utah State University, after which he worked as a thermal designer/analyst in the aerospace and electronics industries. After several years of engineering practice, he resumed his formal education, earning his Ph.D. in mechanical engineering at the University of Utah. Hagen is the author of an undergraduate heat transfer text. Having drawn upon his industrial and

teaching experience, he strongly believes that engineering students must develop effective analytical problem solving abilities. His book, *Introduction to Engineering Analysis*, was written to help beginning engineering students learn a systematic approach to engineering analysis.

Mark N. Horenstein is a Professor in the Department of Electrical and Computer Engineering at Boston University. He has degrees in Electrical Engineering from M.I.T. and U.C. Berkeley and has been involved in teaching engineering design for the greater part of his academic career. He devised and developed the senior design project class taken by all electrical and computer engineering students at Boston University. In this class, the students work for a virtual engineering company developing products and systems for real-world engineering and social-service clients. Many of the design projects developed in his class have been aimed at assistive technologies for individuals with disabilities.

Acknowledgments: I would like to thank Prof. James Bethune, the architect of the Peak Performance event at Boston University, for his permission to highlight the competition in my text. Several of the ideas relating to brainstorming and teamwork were derived from a workshop on engineering design offered by Prof. Charles Lovas of Southern Methodist University. The principles of estimation were derived in part from a freshman engineering problem posed by Prof. Thomas Kincaid of Boston University.

Scott D. James is a staff lecturer at Kettering University (formerly GMI Engineering & Management Institute) in Flint, Michigan. He is currently pursuing a Ph.D. in Systems Engineering with an emphasis on software engineering and computer-integrated manufacturing. Scott decided on writing textbooks after he found a void in the books that were available. "I really wanted a book that showed how to do things in good detail but in a clear and concise way. Many of the books on the market are full of fluff and force you to dig out the really important facts." Scott decided on teaching as a profession after several years in the computer industry. "I thought that it was really important to know what it was like outside of academia. I wanted to provide students with classes that were up to date and provide the information that is really used and needed."

Acknowledgments: Scott would like to acknowledge his family for the time to work on the text and his students and peers at Kettering who offered helpful critiques of the materials that eventually became the book.

David C. Kuncicky is a native Floridian. He earned his Baccalaureate in psychology, Master's in computer science, and Ph.D. in computer science from Florida State University. He has served as a faculty member in the Department of Electrical Engineering at the FAMU–FSU College of Engineering and the Department of Computer Science at Florida State University. He has taught computer science and computer engineering courses for over 15 years. He has published research in the areas of intelligent hybrid systems and neural networks. He is currently the Director of Engineering at Bioreason, Inc. in Sante Fe, New Mexico.

Acknowledgments: Thanks to Steffie and Helen for putting up with my late nights and long weekends at the computer. Thanks also to the helpful and insightful technical reviews by Jerry Ralya, Kathy Kitto, Avi Singhal, Thomas Hill, Ron Eaglin, Larry Richards, and Susan Freeman. I appreciate the patience of Eric Svendsen and Joe Russo of Prentice Hall for gently guiding me through this project. Finally, thanks to Susan Bassett for having faith in my abilities, and for providing continued tutelage and support.

Ron Larsen is an Associate Professor of Chemical Engineering at Montana State University, and received his Ph.D. from the Pennsylvania State University. He was initially attracted to engineering by the challenges the profession offers, but also appreciates that engineering is a serving profession. Some of the greatest challenges he has faced while teaching have involved non-traditional teaching methods, including evening courses for practicing engineers and teaching through an interpreter at the Mongolian National University. These experiences have provided tremendous opportunities to learn new ways to communicate technical material. He tries to incorporate the skills he has learned in non-traditional arenas to improve his lectures, written materials, and learning programs. Dr. Larsen views modern software as one of the new tools that will radically alter the way engineers work, and his book *Introduction to Mathcad* was written to help young engineers prepare to meet the challenges of an ever-changing workplace.

Acknowledgments: To my students at Montana State University who have endured the rough drafts and typos, and who still allow me to experiment with their classes—my sincere thanks.

Sanford Leestma is a Professor of Mathematics and Computer Science at Calvin College, and received his Ph.D. from New Mexico State University. He has been the long-time co-author of successful textbooks on Fortran, Pascal, and data structures in Pascal. His current research interest are in the areas of algorithms and numerical computation.

Jack Leifer is an Assistant Professor in the Department of Mechanical Engineering at the University of Kentucky Extended Campus Program in Paducah, and was previously with the Department of Mathematical Sciences and Engineering at the University of South Carolina—Aiken. He received his

Ph.D. in Mechanical Engineering from the University of Texas at Austin in December 1995. His current research interests include the modeling of sensors for manufacturing, and the use of Artificial Neural Networks to predict corrosion.
Acknowledgements: I'd like to thank my colleagues at USC—Aiken, especially Professors Mike May and Laurene Fausett, for their encouragement and feedback; Eric Svendsen and Joe Russo of Prentice Hall, for their useful suggestions and flexibility with deadlines; and my parents, Felice and Morton Leifer, for being there and providing support (as always) as I completed this book.

Richard M. Lueptow is the Charles Deering McCormick Professor of Teaching Excellence and Associate Professor of Mechanical Engineering at Northwestern University. He is a native of Wisconsin and received his doctorate from the Massachusetts Institute of

Technology in 1986. He teaches design, fluid mechanics, and spectral analysis techniques. "In my design class I saw a need for a self-paced tutorial for my students to learn CAD software quickly and easily. I worked with several students a few years ago to develop just this type of tutorial, which has since evolved into a book. My goal is to introduce students to engineering graphics and CAD, while showing them how much fun it can be." Rich has an active research program on rotating filtration, Taylor Couette flow, granular flow, fire suppression, and acoustics. He has five patents and over 40 refereed journal and proceedings papers along with many other articles, abstracts, and presentations.
Acknowledgments: Thanks to my talented and hard-working co-authors as well as the many colleagues and students who took the tutorial for a "test drive." Special thanks to Mike Minbiole for his major contributions to Graphics Concepts with SolidWorks. Thanks also to Northwestern University for the time to work on a

book. Most of all, thanks to my loving wife, Maiya, and my children, Hannah and Kyle, for supporting me in this endeavor. (Photo courtesy of Evanston Photographic Studios, Inc.)

Larry Nyhoff is a Professor of Mathematics and Computer Science at Calvin College. After doing bachelor's work at Calvin, and Master's work at Michigan, he received a Ph.D. from Michigan State and also did graduate work in computer science at Western Michigan. Dr. Nyhoff

has taught at Calvin for the past 34 years—mathematics at first and computer science for the past several years. He has co-authored several computer science textbooks since 1981 including titles on Fortran and C++, as well as a brand new title on Data Structures in C++.
Acknowledgments: We express our sincere appreciation to all who helped in the preparation of this module, especially our acquisitions editor Alan Apt, managing editor Laura Steele, developmental editor Sandra Chavez, and production editor Judy Winthrop. We also thank Larry Genalo for several examples and exercises and Erin Fulp for the Internet address application in Chapter 10. We appreciate the insightful review provided by Bart Childs. We thank our families—Shar, Jeff, Dawn, Rebecca, Megan, Sara, Greg, Julie, Joshua, Derek, Tom, Joan; Marge, Michelle, Sandy, Lory, Michael—for being patient and understanding. We thank God for allowing us to write this text.

Paul Riley is an author, instructor, and designer specializing in graphics and design for multimedia. He is a founding partner of CAD Support Associates, a contract service and professional training organization for computer-aided design. His 15 years of business experience and 20 years of teach-

ing experience are supported by degrees in education and computer science. Paul has taught AutoCAD at the University of Massachusetts at Lowell and is presently teaching AutoCAD at Mt. Ida College in Newton, Massachusetts. He has developed a program, Computer-aided Design for Professionals that is highly regarded by corporate clients and has been an ongoing success since 1982.

Dr. Robert Rizza is an Assistant Professor of Mechanical Engineering at North Dakota State University, where he teaches courses in mechanics and computer-aided design. A native of Chicago, he received the Ph.D. degree from the Illinois Institute of Technology. He is also the author of Getting Starting

with Pro/Engineer. Dr. Rizza has worked on a diverse range of engineering projects including projects from the railroad, bioengineering, and aerospace industries. His current research interests include the fracture of composite materials, repair of cracked aircraft components, and loosening of prostheses.

Peter Schiavone is a professor and student advisor in the Department of Mechanical Engineering at the University of Alberta, Canada. He received his Ph.D. from the University of Strathclyde, U.K. in 1988. He has authored several books in the area of student academic success as well as numerous papers in international scientific research journals. Dr. Schiavone has worked in private industry in several different areas of engineering including aerospace and systems engineering. He founded the first Mathematics Resource Center at the University of Alberta, a unit designed specifically to teach new students the necessary *survival skills* in mathematics and the physical sciences required for success in first-year engineering. This led to the Students' Union Gold Key Award for outstanding contributions to the university. Dr. Schiavone lectures regularly to freshman engineering students and to new engineering professors on engineering success, in particular about maximizing students' academic performance. He wrote the book *Engineering Success* in order to share the *secrets of success in engineering study*: the most effective, tried and tested methods used by the most successful engineering students.

Acknowledgements: Thanks to Eric Svendsen for his encouragement and support; to Richard Felder for being such an inspiration; to my wife Linda for sharing my dreams and believing in me; and to Francesca and Antonio for putting up with Dad when working on the text.

David I. Schneider holds an A.B. degree from Oberlin College and a Ph.D. degree in Mathematics from MIT. He has taught for 34 years, primarily at the University of Maryland. Dr. Schneider has authored 28 books, with one-half of them computer programming books. He has developed three customized software packages that are supplied as supplements to over 55 mathematics textbooks. His involvement with computers dates back to 1962, when he programmed a special purpose computer at MIT's Lincoln Laboratory to correct errors in a communications system.

David I. Schwartz is an Assistant Professor in the Computer Science Department at Cornell University and earned his B.S., M.S., and Ph.D. degrees in Civil Engineering from State University of New York at Buffalo. Throughout his graduate studies, Schwartz combined principles of computer science to applications of civil engineering. He became interested in helping students learn how to apply software tools for solving a variety of engineering problems. He teaches his students to learn incrementally and practice frequently to gain the maturity to tackle other subjects. In his spare time, Schwartz plays drums in a variety of bands.

Acknowledgments: I dedicate my books to my family, friends, and students who all helped in so many ways. Many thanks go to the schools of Civil Engineering and Engineering & Applied Science at State University of New York at Buffalo where I originally developed and tested my UNIX and Maple books. I greatly appreciate the opportunity to explore my goals and all the help from everyone at the Computer Science Department at Cornell. Eric Svendsen and everyone at Prentice Hall also deserve my gratitude for helping to make these books a reality. Many thanks, also, to those who submitted interviews and images.

Michael T. Snyder is President of Internet startup Appointments123.com. He is a native of Chicago, and he received his Bachelor of Science degree in Mechanical Engineering from the University of Notre Dame. Mike also graduated with honors from Northwestern University's Kellogg Graduate School of Management in 1999 with his Masters of Management degree. Before Appointments123.com, Mike was a mechanical engineer in new product development for Motorola Cellular and Acco Office Products. He has received four patents for his mechanical design work. "Pro/Engineer was an invaluable design tool for me, and I am glad to help students learn the basics of Pro/Engineer."

Acknowledgments: Thanks to Rich Lueptow and Jim Steger for inviting me to be a part of this great project. Of course, thanks to my wife Gretchen for her support in my various projects.

Jim Steger is currently Chief Technical Officer and cofounder of an Internet applications company. He graduated with a Bachelor of Science degree in Mechanical Engineering from Northwestern University. His prior work included mechanical engineering assignments at Motorola and Acco Brands. At Motorola, Jim worked on part design for two-way radios and was one of the lead mechanical engineers on a cellular phone product line. At Acco Brands, Jim was the sole engineer on numerous office product designs. His Worx stapler has won design awards in the United States and in Europe. Jim has been a Pro/Engineer user for over six years.

Acknowledgments: Many thanks to my co-authors, especially Rich Lueptow for his leadership on this project. I would also like to thank my family for their continuous support.

Reviewers

ESource benefited from a wealth of reviewers who on the series from its initial idea stage to its completion. Reviewers read manuscripts and contributed insightful comments that helped the authors write great books. We would like to thank everyone who helped us with this project.

Concept Document

Naeem Abdurrahman *University of Texas, Austin*
Grant Baker *University of Alaska, Anchorage*
Betty Barr *University of Houston*
William Beckwith *Clemson University*
Ramzi Bualuan *University of Notre Dame*
Dale Calkins *University of Washington*
Arthur Clausing *University of Illinois at Urbana —Champaign*
John Glover *University of Houston*
A.S. Hodel *Auburn University*
Denise Jackson *University of Tennessee, Knoxville*
Kathleen Kitto *Western Washington University*
Terry Kohutek *Texas A&M University*
Larry Richards *University of Virginia*
Avi Singhal *Arizona State University*
Joseph Wujek *University of California, Berkeley*
Mandochehr Zoghi *University of Dayton*

Books

Stephen Allan *Utah State University*
Naeem Abdurrahman *University of Texas, Austin*
Anil Bajaj *Purdue University*
Grant Baker *University of Alaska—Anchorage*
Betty Burr *University of Houston*
William Beckwith *Clemson University*
Haym Benaroya *Rutgers University*
Tom Bledsaw *ITT Technical Institute*
Tom Bryson *University of Missouri, Rolla*
Ramzi Bualuan *University of Notre Dame*
Dan Budny *Purdue University*
Dale Calkins *University of Washington*
Arthur Clausing *University of Illinois*
James Devine *University of South Florida*

Patrick Fitzhorn *Colorado State University*
Dale Elifrits *University of Missouri, Rolla*
Frank Gerlitz *Washtenaw College*
John Glover *University of Houston*
John Graham *University of North Carolina—Charlotte*
Malcom Heimer *Florida International University*
A.S. Hodel *Auburn University*
Vern Johnson *University of Arizona*
Kathleen Kitto *Western Washington University*
Robert Montgomery *Purdue University*
Mark Nagurka *Marquette University*
Romarathnam Narasimhan *University of Miami*
Larry Richards *University of Virginia*
Marc H. Richman *Brown University*
Avi Singhal *Arizona State University*
Tim Sykes *Houston Community College*
Thomas Hill *SUNY at Buffalo*
Michael S. Wells *Tennessee Tech University*
Joseph Wujek *University of California, Berkeley*
Edward Young *University of South Carolina*
Mandochehr Zoghi *University of Dayton*
John Biddle *California State Polytechnic University*
Fred Boadu *Duke University*
Harish Cherukuri *University of North Carolina —Charlotte*
Barry Crittendon *Virginia Polytechnic and State University*
Ron Eaglin *University of Central Florida*
Susan Freeman *Northeastern University*
Frank Gerlitz *Washtenaw Community College*
Otto Gygax *Oregon State University*
Donald Herling *Oregon State University*
James N. Jensen *SUNY at Buffalo*
Autar Kaw *University of South Florida*
Kenneth Klika *University of Akron*
Terry L. Kohutek *Texas A&M University*
Melvin J. Maron *University of Louisville*
Soronadi Nnaji *Florida A&M University*
Michael Peshkin *Northwestern University*
Randy Shih *Oregon Institute of Technology*
Neil R. Thompson *University of Waterloo*
Garry Young *Oklahoma State University*

Contents

1

Mechanical Engineering as a Profession

1.1 INTRODUCTION

What exactly is a mechanical engineer? What does a mechanical engineer do? How does one go about becoming a mechanical engineer? These are some of the questions pertaining to mechanical engineers and their role in society that we will answer in this chapter.

For introduction, we may say that mechanical engineers work in many industries, often developing everyday products. Think for a moment about some of the products that you may have used. Did you get a drink of water, take a shower, drive a car, or ride a bike? The products used to perform all these activities involved a mechanical engineer in one way or another.

In developing these and other commercial products, mechanical engineers often work on a team, whose members may be from other disciplines. Thus, mechanical engineers interact with other professionals whose experience may lie in other engineering fields or in finance and marketing.

This is just a brief synopsis of the role of a mechanical engineer. In the sections that follow, we will examine this role further.

1.2 THE ROLE OF A MECHANICAL ENGINEER

Mechanical engineers are professionals devoted to employing the principles of motion, forces, and energy. Machines may be used to convert one motion to another, transform energy from one form to another, or apply forces. This implies that mechanical engineers are very much interested in the design, analysis, and fabrication of machinery.

OBJECTIVES

After reading this chapter, you should be able to do the following:

- Understand the role of mechanical engineering on society and the impact that this role has on everyday life.
- Recognize that mechanical engineers are problem solvers who approach problems in a logical way.
- Use a logical approach to solving problems.

To accomplish these tasks, a mechanical engineer must have a firm foundation in the underlying engineering sciences. These sciences include the study of the motion of fluids and gases, the deformation of solid materials, and the study of materials. To understand and apply principles from these sciences, mechanical engineers must have a firm grounding in science and mathematics.

To accomplish their goals, mechanical engineers work in a team environment, often with professionals from other disciplines. Because machines are used in numerous applications, mechanical engineers are employed in many different fields. Power generation, aerospace technology, and automotives are just a few industries that employ mechanical engineers. In general, mechanical engineers work in industry, developing various commercial products, and also work at universities, in government, and in consulting firms.

Mechanical engineers employed at a university teach the next generation of engineers. These individuals are motivated by their love of teaching and their desire to convey their experiences and knowledge to others. They also direct research activities and write books and technical papers.

Those working in government are employed at various research centers and help to focus the national attention towards new and promising areas of technology. NASA is a government agency that employs many mechanical engineers that develop technology for the aerospace sciences.

PROFESSIONAL SUCCESS

At NASA, the focus is divided into four areas called enterprises: the Office of AeroSpace Technology, Human Exploration, and Development of Space (HEDS), Destination: Earth and the Office of Space Science. NASA centers work on various projects that fit in to these enterprises. For example, one of the goals of the Office of AeroSpace Technology is to develop an affordable high-speed transport that cuts the travel time to the far east and Europe by 50%, while meeting stringent aircraft noise and emission standards.

One of the endeavors in attaining these goals is to implement new advanced low-cost material and structural concepts. Mechanical engineers work side by side with material scientists developing and implementing new materials in structural applications. Because the aircraft will have to travel at very high speeds, the heating of certain portions of the exterior of the aircraft is a problem. Engineers use techniques developed from the study of thermal sciences and materials to dissipate or deal with this heat.

Mechanical engineers employed in consulting firms use their expertise to solve specific problems. Many mechanical engineers start consulting firms or join consulting firms after many years of experience.

Mechanical engineers work side by side with sales and marketing professionals. But, because of today's global marketplace and the needs of their employers, many mechanical engineers are being trained to understand the principles of marketing and sales, so that those principles can be incorporated into the design of the product.

Mechanical engineers are also finding employment in nontraditional engineering areas such as the environment and medicine. Mechanical engineers are now employed to design equipment to clean and preserve the environment and work with natural scientists such as biologists. Mechanical engineers are also employed in the Bioengineering industry, along with medical doctors, physical therapists, and other medical professionals in endeavors to analyze the human body, design and fabricate replacement limbs, and design and construct medical instrumentation.

As can be seen from the preceding paragraphs, mechanical engineers play an important role in society and find employment in a wide range of areas. Thus, it is very difficult to find an area of interest where mechanical engineers have not played an active role.

PROFESSIONAL SUCCESS

Mechanical engineers have a responsibility to society to design and build safe, reliable, and efficient machinery. Also, mechanical engineers need to consider what long-term impacts their designs have on the environment.

One area of concern is pollution. Global warming due to the release of green house gases is a hotly debated topic nowadays. However, there is evidence that the industrial age has resulted in polluted landscapes, air, and watersheds. With increasing emphasis on this topic, it is quite possible that endeavors to minimize or eliminate pollution may be a focus of your career.

The exhaust from the internal combustion automobile engine has been blamed for much of the release of greenhouse gases. Some efforts in the last 20 years have concentrated on reformulating gasoline so that it burns cleaner. In addition, mechanical engineers working on new automobile designs have developed lighter cars, which are more fuel efficient. They have also developed and are continuing to improve engines that run more efficiently and therefore pollute less.

1.3   BECOMING A MECHANICAL ENGINEER AND THE LIFELONG LEARNING PROCESS

The mechanical engineer has his or her roots in the millwright who worked in early industrial age industries, such as iron smelters, forges, and textile mills. Millwrights did not have any formal education. However, today, engineering practice requires that mechanical engineers attain formal training.

Formal training for a mechanical engineer is usually attained at a four-year accredited college or university. For the first two years, mechanical engineering students study calculus, chemistry, physics, statics, and dynamics. Many students take the courses covering this material at a two-year college and then transfer to a four-year university. In the last two years of study, students study specialized mechanical engineering topics in machinery, computer-aided design (CAD), computer-aided manufacturing (CAM), thermodynamics, fluid dynamics, materials, and manufacturing processes.

After four years of successful study, the graduate attains a bachelor of science (B.S.). Upon graduation, many students pursue graduate studies and obtain a master of science (M.S.) and a doctor of philosophy (Ph.D.). Some industries, such as the aerospace industry, have traditionally desired engineers with advanced degrees. Many companies are encouraging their employees to pursue graduate studies, and often they pay for their employees' tuition.

Learning for a mechanical engineer does not stop upon graduation, whatever the degree. Learning is a lifelong experience. Because mechanical engineers find employment in areas stressing technology and technology changes over time, it is important that the mechanical engineer continue the educational process throughout his or her career.

Beyond the formal education at a university, additional education occurs from the first day the mechanical engineer steps into his or her place of employment. This may involve specific methods used by the employer. Quite often, the employer will train the employee in areas that are too focused for study at the university level, but necessary in the job.

Mechanical engineering societies also play a role in the continued learning process by offering training seminars and opportunities to meet other engineers. The American Society of Mechanical Engineers (ASME) is the primary professional organization, but mechanical engineers also belong to other societies. Some of these societies are listed in Table 1-1 along with their Internet Web addresses.

0

Many universities will have student chapters of these societies. Participation in one or more of these societies is encouraged, because it helps to build ties with the profession. The societies often sponsor student design competitions, which can be very rewarding and fun. The cost for a student to join one of the societies is usually very low.

TABLE 1-1 Web Site Addresses for Organizations with Mechanical Engineer Members

ORGANIZATION	ADDRESS
American Astrononautical Society	www.aas.org
American Institute of Aeronautics and Astronautics	www.aiaa.org
American Society for Testing and Materials	www.astm.org
American Society of Heating, Refrigeration, and Air-Conditioning Engineers	www.ashrae.org
American Society of Mechanical Engineers	www.asme.org
Society of Automotive Engineers	www.sae.org
Society of Manufacturing Engineers	www.sme.org
Society of Petroleum Engineers	www.spe.org

1.4   APPROACHING AND SOLVING AN ENGINEERING PROBLEM

Mechanical engineers are problem solvers. The mechanical engineer uses techniques honed during training to evaluate a physical system, model the system using known principles, and solve for the unknowns. To be a successful mechanical engineer, you must develop these problem-solving skills early in your academic career.

Students tend to have difficulty with word problems. The following sequences of steps may be used to approach and solve such problems:

1. Read through the problem statement at least twice.
2. Identify the known and unknown quantities. The known quantities are used to solve the problem for the unknown quantities.
3. Sketch a diagram of the problem. This will help you to see what the problem is all about.
4. Translate the words in the problem to a mathematical statement. In doing this, look for key words that convey the concepts of the problem. For example, "twice b" implies $2b$; "is," "was," and "becomes" all mean " $=$."
5. Solve for the unknowns using the appropriate mathematical techniques.
6. Check your answer to see if it is mathematically correct.

EXAMPLE 1.1:   What number added to nine is four times the number?

SOLUTION

Step 2
The problem is actually looking for an unknown number. Let this number be called x.

Step 4
We determine the necessary equation by recognizing that "added to" means " + ," "is" means " = ," and "times" means " × ." Thus, we have

$$x + 9 = 4x.$$

Step 5
We can solve this equation using simple algebra:

$$x + 9 = 4x.$$

$$(1 - 4)x = -9$$

$$\therefore x = 3.$$

Step 6
We check our result by substituting the number back into the original expression:

$$3 + 9 = 4(3)$$

$$12 = 12.$$

PRACTICE!   Find two numbers such that their difference is five and whose sum is also five.

PROFESSIONAL SUCCESS

A good engineer always checks to see if the results obtained from a calculation makes sense. This goes beyond checking for calculation errors by substituting the number back in to the original expression, as was done in step 6 of Example 1.1. It also involves considering whether the result makes physical sense. This is especially true when using complicated computer programs such as finite-element programs to solve for unknown quantities. For example, suppose that a study is undertaken to deter- mine the loads on the structure of a new building due to a heavy snowfall. One would expect that a large amount of snow on the roof of the building would cause the struc- ture of the building to be compressed. Now suppose that after running a computer program you find that the net result is for the building to extend in height! Obviously, something is wrong and the results obtained are impracti- cal at best. At this point, you would need to discard the solution and investigate exactly what went wrong.

1.5   SUMMARY

Studying at an accredited college or university helps the student attain formal training to become a mechanical engineer. The topics studied range from math and science to specialized courses in mechanical engineering. Many students, upon graduation, seek advanced degrees.

Whatever the degree, learning for a mechanical engineer does not stop upon graduation. Learning is a lifelong experience. Often, a mechanical engineer is trained on the job with respect to special tasks necessary to complete his or her duties. Also, because mechanical engineers work with technology that is ever changing, the success- ful engineer must keep abreast of the current state of technology.

Mechanical engineers are problem solvers who are employed in a wide variety of areas. They work in a team environment, often with professionals from other disciplines. Mechanical engineers work in many industries, often developing everyday products. They are also employed in government, doing research, and in academia, teaching the next generation of engineers. Thus, mechanical engineers play an important role in society.

KEY TERMS   Mechanical engineer   lifelong learning   solving problems

1.6   EXERCISES

For problems 1.1 through 1.6, ask an instructor or a mechanical engineer to discuss the topics presented. Then, for each topic, write a paragraph that includes the ideas discussed by the professional, detailing what role each topic plays in mechanical engineering. Also, include comments on the types of engineering problems that the engineer might face and have to solve related to the topic.

E1.   How are principles from fluid mechanics used in the design of automobiles to reduce fuel consumption?

E2.   How is the theory of solid mechanics used to design structural components such as automobile and aircraft panels? What are some of the assumptions in the development of this theory? How do these assumptions influence analysis and design of these structural components?

E3.   What principles from the theory of thermodynamics are used to design fossil-fuel power generation plants? What parameters from the theory of thermodynamics influence the efficiency of these plants?

E4.   How is the theory heat transfer used to design mechanisms to efficiently remove heat from small engines?

E5.   What mechanisms are used in the design of a recliner? What are the principles of dynamics that are used in the design of these mechanisms?

E6.   What graphical techniques are used to communicate ideas in engineering design? How does computer-aided design (CAD) fit in with engineering design?

For Problems 1.7 through 1.10, use the steps in Section 1.4 to solve the word problems.

E7.   Find a number whose triple is eight more than the result of adding the original number plus five.

E8.   Find two consecutive integers whose sum is three times the smaller number.

E9.   What number added to 10 is the same as one-third the number.

E10.   A box is to be constructed from cardboard so that each face is a square. Suppose that the total surface area of the box is 2000 cm². What is the size of each face?

2

Dimensions, Units, and Error

2.1   INTRODUCTION

The engineering sciences depend on the quantification of physical phenomena. That is, in order for an engineer to make use of a physical quantity, it must have a numerical value. This numerical value is often obtained through measurement. A unit of measurement is associated with the parameter.

Because measurement instruments are imperfect, there will be some error in the experimentally obtained value. It is important for an engineer to have some indication of the size of this error.

In this chapter, we will examine the systems of measurements commonly used by mechanical engineers. Furthermore, we discuss the types of error that occur whenever a measurement is made.

2.2   DIMENSIONS AND UNITS

In engineering practice, physical quantities are quantified by giving a magnitude and a unit of measurement. For example, the playing surface of a football field is one hundred yards long. In this case, the magnitude is the value one hundred, and the yard is the unit of measurement. A *unit* refers to the arbitrary system of measurement.

Units of measurement are standardized quantities. In the United States, the National Institute of Standards and Technology (NIST) standardizes units of measurement. Standardization of a unit means that within the region where the standardizing organization has jurisdiction, the unit has the same meaning for everyone. For example, the yard in the previous example would imply the same physical quantity of length to a person living in California as it would

OBJECTIVES

After reading this chapter, you should be able to do the following:

- Understand the difference between a unit and a dimension.
- Convert between the different systems of measurement.
- Understand the types of error in a measurement.
- Express large numbers in scientific notation.

to a person living in New York. Furthermore, a standard yard would have the same length for both individuals, because the length of a yard is standardized.

Some units are combinations of others. For instance, the unit of force is a combination of the units for mass and acceleration.

A *dimension* is a measurable quantity of a physical parameter. The yard is a unit of length in a particular system of measurement whose dimension is length. *Fundamental dimensions* are independent of one another. For example, a set of fundamental dimensions may be mass, length, time, and temperature.

Let us define the dimension of an engineering quantity by placing a bracket around the quantity. Thus, if a is acceleration, then $[a]$ would imply the dimensions of acceleration. Furthermore, let M, L, T and θ represent the fundamental dimensions of mass, length, time, and temperature, respectively. Then, the dimensions of acceleration would be

$$[a] = \frac{[L]}{[T^2]} = \frac{L}{T^2}.$$

Therefore, the dimensions of acceleration are length per time squared. The dimension of mass is mass. What are the dimensions of force, F?

The law of *Dimensional Homogeneity* requires that all terms in an equation have the same dimensions. Newton's second law says that the force on a system is equal to the mass of the system times the acceleration. This may be written as $F = ma$, where F is the force, m is the mass, and a is the acceleration. Therefore, the dimensions of force must be the same as those of mass times the acceleration in order for Newton's second law to be valid. Then, the dimensions of force are

$$[F] = [ma] = [m][a] = m\frac{L}{T^2}.$$

A nondimensional quantity is a quantity that has dimensions of one. For example, what are the dimensions of $F(ma)$? The answer is

$$\left[\frac{F}{ma}\right] = \frac{\left[\dfrac{mL}{T^2}\right]}{\left[m\dfrac{L}{T^2}\right]} = 1.$$

Nondimensional quantities play an important role in fluid mechanics.

PRACTICE!

You may have noticed that when you pull on a spring, there is a force tending to restore the spring to it's unstretched position. If F is this force and x is the amount that the spring was displaced, engineers write the relationship between the force and spring as

$F = kx,$

where k is a constant called the spring stiffness. What must the dimensions of k be for this equation to obey the law of dimensional homogeneity?

2.2.1   The British Gravitational System

The units of a quantity depend on the arbitrary system of measurement. One of these systems of measurement is the British Gravitational System (BGS). This particular system uses force, as a fundamental dimension, instead of mass. Because of Newton's second law (i.e., $F = ma$), the dimensions of mass become

$$[m] = \frac{[F]}{[a]} = \frac{FT^2}{L}.$$

In BGS, the unit of force is the pound (lb). Therefore, the unit of mass is related to the unit pound. This unit is called the *slug* and is defined by

$$1 \text{ slug} = 1 \text{ lb} \cdot \text{sec}^2/\text{ft},$$

where ft (foot) is the unit of length.

The weight of an object, W, is the mass times the gravitational acceleration constant, g. This may be written as $W = mg$. In this case, the unit is the pound, as it is a force. Because, on earth, $g = 32.174 \text{ ft/sec}^2$, a slug weighs

$$\text{Weight of slug} = W = 1 \quad \text{slug} \cdot 32.174 \text{ ft/sec}^2 = 32.174 \text{ lb}. \tag{2-1}$$

Therefore, in BGS, one slug has a weight of 32.174 pounds.

Additional units for BGS are listed in Table 2-1.

TABLE 2-1   Listed in the Table are Some Common Units for Engineering Parameters. The Parameter in Parentheses is the Symbol for That Unit

QUANTITY	BGS	EE	SI
Mass	slug	lbm	kilogram
Length	ft	ft	meter (m)
Time	sec (s)	sec (s)	sec (s)
Area	ft²	ft²	m²
Volume	ft³	ft³	m³
Velocity	Ft/s	Ft/s	M/s
Acceleration	Ft/s²	Ft/s²	M/s²
Density	Slug/ft³	Lbm/ft³	Kg/m³
Force	Lbf	Lbf	newton (N) = kg·m/s²
Pressure (or Stress)	Lbf/ft²	Lbf/ft²	pascal (Pa) = N/m²
Energy (or Work)	ft·Lbf	ft·Lbf	joule (J) = N·m
Power	ft·Lbf/s	ft·Lbf/s	watt (W) = J/s

2.2.2   The English Engineering System

The English Engineering (EE) system is defined so that 1 unit of mass has a weight of one pound (Lbf) in a standard gravity. The unit of mass in the EE system is the pound-mass (Lbm), and the unit of force is called the pound-force (Lbf). From Newton's second law, we have that $1 \text{ Lbf} = 1 \text{ Lbm} \cdot 32.174 \text{ ft/sec}^2$. This relation does not have the same units on the right-hand side as the left-hand side of the equation. In order not to violate the law of dimensional homogeneity, the pound-force must be related to the unit of pound mass through the relation:

$$1 \text{ Lbf} = \frac{1 \text{ Lbm} \cdot 32.174 \text{ ft/sec}^2}{32.174 \text{ ft} \cdot \text{Lbm/Lbf} \cdot \text{sec}^2}. \tag{2-2}$$

Thus, in the EE system, when converting between units of mass or force, one must divide or multiply by a conversion factor, g_c, equal to 32.174 ft·Lbm/Lbf·sec². Then, in order to have dimensional homogeneity, Newton's second law is properly written as

$$F = \frac{ma}{g_c}. \tag{2-3}$$

EXAMPLE 2.1:   What is the force (in Lbf) on a 0.1-Lbm body accelerating at 2 ft/sec²?

SOLUTION

The force is given by Equation 2-3. Therefore, we have

$$F = \frac{0.1 \text{ Lbm} \cdot 32.174 \text{ ft/sec}^2}{32.174 \text{ ft} \cdot \text{Lbm/Lbf} \cdot \text{sec}^2} = 0.1 \text{ Lbf}$$

Additional units for the EE system are listed in Table 2-1.

2.2.3   The SI System

The Système International (SI), or metric system, was first established in France in about 1800 and has become the worldwide standard. In the United States, voluntary conversion to the metric system began in 1975, even though the system has been recognized by Congress as early as 1866. In time, it is expected that all units will be in SI form.

The unit of mass in SI is the kilogram (kg). The unit of length is the meter (m) and for volume, the liter (L). The system is set up so that one liter of water has a mass of one kilogram. The unit of weight in the SI system is the newton (N). From newton's second law, one newton is defined as

$$1 \text{ N} = 1 \text{ kg} \cdot \text{m/sec}^2.$$

The weight is given by the expression $W = mg$, where g is 9.81 m/sec².

A clear advantage of SI over other systems of measurement is that SI is based on the powers of 10. For example, 1/100th of a meter is a centimeter (cm). To convert from meter to centimeter, all one need do is multiply by 100. This is easily memorized, because "centi" means 1/100th. Therefore, centimeter means 1/100th of a meter. Table 2-2 lists the prefixes defined in the SI system.

TABLE 2-2   Prefixes for the SI System

MULTIPLICATION FACTOR	PREFIX	SI SYMBOL
10^{12}	tera	T
10^9	giga	G
10^6	mega	M
10^3	kilo	k
10^2	hecto	h
10^1	deka	da
10^{-1}	deci	d
10^{-2}	centi	c
10^{-3}	milli	m

TABLE 2-2   Prefixes for the SI System

MULTIPLICATION FACTOR	PREFIX	SI SYMBOL
10^{-6}	micro	μ
10^{-9}	nano	n
10^{-12}	pico	p
10^{-15}	femto	f
10^{-18}	atto	a

Handling small quantities is very easy in SI. Consider a measurement of 0.03 m. Rather than writing this number in decimal form, multiply it by 100 to get the same measurement in centimeters (i.e., 3 cm). Conversion between SI and the BGS and EE systems is discussed in Section 2.3.

2.3   CONVERSION BETWEEN DIFFERENT UNITS

Table 2-3 contains a table with conversion factors. It is not an exhaustive table, but may be used for the exercises in this book.

TABLE 2-3   The table lists conversion factors that may be used to solve engineering problems. Some values are approximate

DIMENSION	CONVERSION FACTOR
length	0.3048 m = 1 ft
area	$1 \text{ m}^2 = 10.764 \text{ ft}^2 = 1550 \text{ in}^2$
Mass	1 kg = 2.2046 lbm
Volume	$1 \text{ m}^3 = 264.2 \text{ gal} = 35.32 \text{ ft}^3$
velocity	1 m/s = 3.2802 ft/s = 2.237 mi/h
Acceleration	$1 \text{ m/s}^2 = 3.2808 \text{ ft/s}^2$
Force	1 N = 0.22481 lbf
Pressure	$1 \text{ kPa} = 0.14504 \text{ lbf/in}^2$
Temperature	$°C = \dfrac{5}{9}(°F - 32°)$

Except for the definition of mass, the BGS and the EE system have identical units. Conversion between these two units of measurement is straightforward. Because 1 Lbf accelerates 1 Lbm at 32.174 ft/sec^2, 32.174 Lbm would be accelerated at 1 ft/sec^2. Therefore,

$$1 \text{ slug} = 32.174 \text{ Lbm.} \tag{2-4}$$

Expression (2.4) may be used to convert between units of Lbm and slug.

In general, a conversion factor exists between equivalent units. For example, 1 ft = 0.3048 m. When using conversion factors, keep track of the units. In your result, all but the desired unit should drop out from the expression.

EXAMPLE 2.2: A structural component has a length of 2 ft 3 inches. What is the corresponding length in centimeters?

SOLUTION

The conversion between feet and meters is given in Table 2-3 (in particular, 1 ft = 0.3048 m). First, we use the fact that there are 12 inches per foot. Thus, 2 ft 3 inches is equivalent to

$$L = 2 \text{ ft} + 3 \text{ inch} \frac{1 \text{ ft}}{12 \text{ inch}} = 2.25 \text{ ft},$$

which allows us to convert to meters. Thus,

$$L = 2.25 \text{ ft} \times \frac{0.3048 \text{ m}}{1 \text{ ft}} \times \frac{100 \text{ cm}}{1 \text{ m}} = 68.58 \text{ cm}.$$

Notice how all but the units of centimeter drop out in the expression.

PROFESSIONAL SUCCESS

As we have seen, there are three commonly used systems of measurement in engineering. Although the metric system is gaining in popularity, BGS and EE system are still used. Be very careful when performing calculations with values associated with different systems of measurement.

2.4 ERRORS AND ACCURACY

The engineering sciences are based on measurements of fundamental parameters, such as temperature, velocity, and so on. These measurements are performed for three basic reasons.

The first reason is to obtain information about an unknown process. A classic example is Robert Hooke, who, in the 17th century, measured the extension of wires due to a weight. Hooke was motivated by a desire to understand the deformation of a solid material. In the course of his experiments, he uncovered the relationship between forces acting on a body and the deformation of the body. Today, Hooke's discoveries are contained within a relation known as Hooke's law, which relates stress and strain.

The second reason is to obtain parameters that cannot be obtained by theoretical methods. An example is the forces acting on an aircraft wing. Because it is expensive to numerically obtain these forces, models of the wing or aircraft are placed in a wind tunnel and experimentally measured.

The third reason is to validate theory. A theory is valid only if it explains experimentally obtained observations. A theory is invalid if it contradicts such experimental observations.

Accuracy is defined as the difference between the true value and the measured value of a given parameter. For example, a beaker of water may be measured with a thermometer, and a value of 10.5 °C may be obtained. Suppose the true value of the temperature is 10 °C. There is obviously a difference between the two values, for reasons that will be discussed shortly. The difference between the two values is 0.5 °C. Thus, we would say that the measured temperature is accurate to within 0.5 °C.

For the reasons outlined previously, quantities in engineering depend on one or more measurements. However, all measurements contain error. Obviously, if the error in the measurement is large compared with the measured quantity, then the result is not very useful. Thus, of great importance in the measurement of fundamental engineering parameters is not whether there is error, but is in obtaining an estimate of the error.

The error in a measurement may be divided into two parts. The first part, *bias error*, occurs whenever a particular measurement instrument is used. It is inherent, due to the design and nature of the measurement instrument.

For example, suppose you wanted to cut an 8-foot 2 × 4 piece of lumber to, say, 92 inches. Because the smallest graduation on the tape measure is 1/16 of an inch, the measurement can never be better than half that value. Thus, every time the tape measure is used, the cut will be at least off by 1/32 of an inch.

The second part of the error is called the *precision error* and occurs in a random manner, having a different value whenever the measurement is performed. Precision error is often the result of mistakes made by the technician or external disturbances, such as temperature variations or vibration in a system.

Repeating a measurement several times using different equipment and technicians will randomize the bias error, provided that enough tests are performed. This effectively transforms the bias error into a type of precision error. Statistical analysis of the results of tests will give a mean value for the desired quantity and a level of confidence in the result.

PRACTICE!

In completing the construction of a house, a finish carpenter uses a table saw to rip five pieces from a 4-foot × 8-foot sheet of oak plywood. The desired width of each piece is four inches. After cutting the five pieces, the carpenter accurately measures each piece and obtains the values in Table 2-4. If the bias error is 1/32 of an inch, what is the precision error for each piece?

TABLE 2-4   Width of the Five Pieces

NUMBER	WIDTH (INCHES)
1	4.010
2	3.995
3	4.120
4	3.998
5	4.005

2.5   SIGNIFICANT DIGITS

Engineering calculations are only accurate to a certain number of significant digits, because the calculations are usually based on one or more measurements. The numbers 4.454 and 0.004454 are both given accurate to four significant figures. The absolute error is less than 0.001 for 4.454 and 0.000001 for 0.004454.

For the value 4.454, the true value lies in the range between 4.4535 and 4.4540. However, for the value 0.004454, the true value lies in the range between 0.0044535 and 0.0044540.

Another way to express a number is by the use of scientific notation. Scientific notation is based on the powers of 10. For example, $10^1 = 10$, $10^2 = 100$, $10^3 = 1,000$, and so on.

Scientific notation is useful in expressing very large or very small numbers. Consider the average distance between the earth and the moon, which is 240,000 miles. In scientific notation, this may be expressed as $240 \times 1,000 = 240 \times 10^3$.

In a number such as 240,000, it is hard to see exactly how many of the digits are significant. Scientific notation may be used to remedy this situation. For example, if five significant figures are required, then 240,000 may be written as 2.40000×10^5. Or, if, say, only three figures are significant, then the number may be written as 240×10^3.

With the advent of calculators, there is a tendency to provide too many figures in a response. Many scientific calculators are available with a nine-digit display. This does not mean that all the figures in the display are significant.

When performing a calculation, remember that the number in the calculation with the least number of significant figures determines the number of significant figures in your answer. For example, consider the calculation:

$$D = 4.25 \text{ ft} \times \frac{0.3048 \text{ m}}{1 \text{ ft}} = 1.30 \text{ m.}$$

In this expression, the number 4.25 has fewer significant figures than 0.3048. Thus, the number of significant figures in 4.25 dictates the number of significant figures in the answer.

The answer has been rounded off. Whenever you round off a number, use the following rule: If the first number discarded is 5 or greater, then increase the preceding number by one.

In the previous example, the result to nine digits was 1.29540000. Because of the number of significant figures in the calculation, this number is rounded off to two decimal places. Because the first number to be discarded is 5, the preceding number (9) is incremented by one. This gives the result 1.30.

PROFESSIONAL SUCCESS

When using scientific notation, be aware of the difference between a number raised to the power of 10 and a number times 10 raised to the power of 10.

For example, note that $1.34^{10} \neq 1.34 \times 10^{10}$, because $1.34^{10} = 18.66$ and $1.34 \times 10^{10} = 13,400,000,000$.

KEY TERMS

accuracy	English Engineering (EE) system	scientific notation
British Gravitational System (BGS)	error	units
dimension	metric system	

2.6   SUMMARY

As we have seen, fundamental quantities in the engineering sciences depend on experimental measurement. These measurements contain error, which is the combination of two types: bias error and precision error. The error in a measurement may be quantified by performing several measurements of the desired quantity. Once the value of the error is quantified, the usefulness of the measurement is known.

Engineering parameters are given in terms of units, which depend on standards codified by governing bodies. A dimension is a measurable quantity of a physical parameter. Dimensions do not depend on the arbitrary system of measurement.

The most widely used system of measurement in the world is SI, or the metric system, although other systems of measurement exist. Once a value of an engineering parameter in known in a particular system of measurement, it may be obtained in another system by using a conversion factor.

2.7   EXERCISES

E1. What is the mass in slugs of an eight-pound bowling ball?

E2. Which has a greater mass, a block of steel weighing 5 pounds or a block of aluminum weighing 10 N?

E3. The speed of light is 299,792,458 m/s. What is the speed of light in km/hr?

E4. Suppose that the steel and aluminum blocks from Problem 2 were each accelerated by a 1-N force. What is the acceleration of the steel block in ft/s²? What is the acceleration of the aluminum block in m/s²?

E5. The gravitational acceleration constant on the moon is 1/6 the value on earth. What will be the weight of a man on the moon if the same man weighs 210 pounds on the Earth?

E6. Suppose that the viscosity of an oil is 0.0003 Lbf·s/ft². What is the viscosity in N·s/m²?

E7. What are the dimensions of the unit of pressure Pa (pascal)? (*Hint:* Consider Table 2-1).

E8. The hydrostatic pressure in a fluid is given by

$$p = p_0 + \rho g y,$$

where p_0 is the pressure at the free surface, ρ is the density of the fluid, y is the depth of the fluid, and g is the acceleration due to gravity. Show that this equation obeys the law of dimensional homogeneity.

E9. A Newtonian fluid is governed by the expression

$$\tau = \mu \frac{\partial u}{\partial y},$$

where τ is the shear stress, having dimensions $m/(LT^2)$ and μ is the viscosity of the fluid, with dimensions $m/(LT)$. What must the units of $\partial u/\partial y$ be in order for this equation to obey the law of dimensional homogeneity?

E10. In conduction, heat is transferred in a body between two different temperatures. The amount of heat transferred per unit area, q'', has units of W/m² (1W = 1J/s = 1N·m/s) in SI. For steady, one dimensional heat transfer along direction x, q'' is given by the expression

$$q'' = k \frac{\Delta T}{\Delta x} \frac{\text{Kelvin}}{\text{meters}}$$

where $\Delta T = T_2 - T_1$ in kelvin, Δx is the change in the dimension x in meters, and k is a constant called the thermal conductivity. Using this relationship and the given information determine the units of the conductivity in SI.

E11. For the equation in Problem E8, what must the units of $\partial u/\partial y$ be if SI is used?

E12. The average distance between the Earth and the Sun is 98 million miles. Write this value in scientific notation with two digits of accuracy.

E13. If the speed of light is 299,792,458 m/s and the distance between the Earth and the Sun is 98 million miles, how long (in minutes) does it take for light to travel from the sun to the Earth?

3

Statics, Dynamics, and Mechanical Engineering

3.1 INTRODUCTION

This chapter deals with a branch of mechanics that is concerned with bodies at rest or in motion acted upon by forces. The bodies in statics and dynamics are treated as rigid bodies; that is, the bodies undergo no deformation. Thus, statics and dynamics form a branch of mechanics called rigid-body mechanics.

Statics deals with bodies that are in equilibrium with applied forces. Such bodies are either at rest or moving at a constant velocity. On the other hand, dynamics deals with bodies that are accelerating. For an accelerating system, the velocity of the system is not constant.

Rigid-body mechanics is based on the Newtons' laws of motion. These laws were postulated for a particle, which has a mass, but no size or shape. Newton's laws may be extended to rigid bodies by considering the rigid body to be made up of a large number of particles whose relative position from each other do not change. Newtons' laws may be stated as follows:

First law. A particle at rest or in motion with constant velocity along a straight line will remain in its present state unless acted upon by an unbalanced force.

Second law. If an unbalanced force acts upon a particle, then the particle will experience an acceleration that has the same direction as the force. The magnitude of the acceleration is proportional to the magnitude of the force.

Third law. For every force acting on a particle, the particle responds with an equal and opposite reactive force.

Rigid-body mechanics is a foundation of mechanical engineering. Because this subject deals with forces applied to a body, a mechanical engineer must be able to obtain a

OBJECTIVES

After reading this chapter, you should be able to do the following:

- Understand the concept of a vector.
- Write a vector in component form.
- Understand the concepts of a couple and moment.
- Construct a free-body diagram of a physical system.

representation of the body with the applied forces. This representation is called a free-body diagram. Furthermore, to predict the response of the body under the applied forces, the engineer must be able to mathematically represent the forces. Such forces may be mathematically represented as vectors.

3.2   THE CONCEPT OF A VECTOR

A *scalar* is a quantity having only a magnitude. For example, the temperature of a cup of coffee is a scalar describing a physical quantity of the coffee.

A *vector* is a quantity having both a magnitude and a direction. Forces are vectors. A force vector directed towards the center of the Earth whose magnitude is the weight of the individual may represent the weight of the person.

In this chapter, using bold typeface will indicate a vector. On the other hand, italic typeface will indicate a scalar. For example, **a** is a vector, whereas *a* is a scalar.

Given two vectors, the vectors will only be equal if both the magnitude and direction of both vectors are equal. A vector that has the same magnitude as another vector, but is in the opposite direction, is a *negative* vector. A vector whose magnitude is unity is called a *unit* vector.

A quantity such as a force has a magnitude, as well as a direction in which it is acting. This implies that vectors may represent forces.

Geometrically, we can represent vectors with an arrow. As shown in Figure 3.1, the head of the arrow points in the direction of the vector, and the length of the arrow indicates the magnitude of the vector.

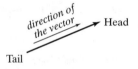

Figure 3.1.

3.2.1   Components of a Vector

If a coordinate system is applied to a physical system, then a vector may be decomposed in vector components that act along each coordinate direction. For example, as shown in Figure 3.2, the vector **F** is the resultant of two components in the two-dimensional Cartesian coordinate system. The vector component $\mathbf{F}_x$ acts along the x direction, while $\mathbf{F}_y$ acts along the y direction. Algebraically, the resultant **F** may be written as

$$\mathbf{F} = \mathbf{F}_x + \mathbf{F}_y. \tag{3-1}$$

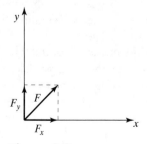

Figure 3.2.

Now, let **i** be a unit vector acting along the x direction and **j** be a unit vector acting along the y direction. The vectors **i** and **j** are sometimes called basis vectors.

Furthermore, suppose that the magnitude of $\mathbf{F}_x$ is F_x and $\mathbf{F}_y$ is F_y, then Equation 3-1 may be written as

$$\mathbf{F} = F_x\mathbf{i} + F_y\mathbf{j}. \tag{3-2}$$

Equation 3-2 treats an expansion for a two-dimensional Cartesian coordinate system. Similar expansions may be obtained for a vector in a non-Cartesian coordinate system.

In general, for a three-dimensional Cartesian coordinate system, a vector **F** may be written as

$$\mathbf{F} = F_x\mathbf{i} + F_y\mathbf{j} + F_z\mathbf{k}, \tag{3-3}$$

where **k** is the unit vector in the z direction.

From geometry (Figure 3.3), it may be shown that the magnitude of a vector is given by

$$\|F\| = \sqrt{F_x^2 + F_y^2 + F_z^2},$$

or the square root of the sum of the squares of the components.

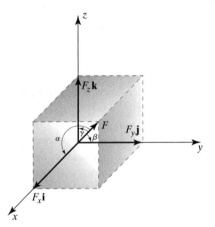

Figure 3.3.

EXAMPLE 3.1:

A boy is pulling on a rope that is tied to a barn door, as shown in Figure 3.4. The boy pulls with a force of 22 N along the direction of the rope, which is at 30° to the vertical. The rope lies in a plane. What are the components of this force?

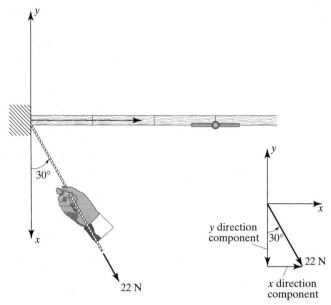

Figure 3.4.

SOLUTION

The problem statement mentions that the 22 N are directed along the direction of the rope. Therefore, we can replace the rope with a force that is located in space in the same manner as the rope. We can place a Cartesian coordinate system where the rope is tied and use geometry to find the components of the force. Because the rope lies in a plane, there are only two components of force, say, F_x and F_y.

The component in the x direction is determined by finding the projection along the x-axis as shown in Figure 3.4. The components of the force along each coordinate axis form a right triangle, as shown in the figure. From trigonometry, the component along the x-axis is

$$F_x = 22 \text{ N} \quad \sin 30 = 11 \text{ N} \rightarrow ,$$

where the arrow indicates the direction of the component.

The component in the y direction is determined by finding the projection along the y-axis, as shown in Figure 3.4. This direction is determined by the cos 30°. Therefore,

$$F_y = 22 \text{ N} \quad \cos 30 = 11\sqrt{3} \text{ N}\downarrow.$$

3.2.2   Direction Cosines and Vectors

Another way to write a vector mathematically is to use cosines of the angles that the vector makes with the Cartesian coordinate directions. These are the angles α, β, and γ shown in Figure 3.3. In the figure, notice that the angles are measured between the tail of the vector **F** and the positive direction of the x-, y-, and z-axes. An interesting fact about these angles is that they are never greater than 180°.

From geometry, it can be shown that the cosine of each angle is equal to the projection along the direction defined by the angle divided by the magnitude of the vector. For example, the cosine of α is found as follows:

$$\cos \alpha = \frac{\text{projection of } \mathbf{F} \text{ on the } x\text{-axis}}{\text{magnitude of } \mathbf{F}}. \tag{3-4}$$

Because the projection of the vector $\mathbf{F}$ on the x-axis is F_x, and the magnitude of $\mathbf{F}$ is $\|F\|$ this becomes

$$\text{Cos } \alpha = \frac{F_x}{\|F\|}. \tag{3-5}$$

Likewise,

$$\text{Cos } \beta = \frac{F_y}{\|F\|} \quad \text{and} \quad \text{Cos } \gamma = \frac{F_z}{\|F\|}. \tag{3-6}$$

The direction cosines are such that the sum of their squares is equal to unity. Thus,

$$\cos^2 \alpha + \cos^2 \beta + \cos^2 \gamma = 1. \tag{3-7}$$

Equation 3-7 implies that only two direction cosines need be obtained from the given information. The third direction cosine may be calculated from the equation.

Now let $\mathbf{C}$ be a vector whose components are the direction cosines; that is, let

$$\mathbf{C} = \cos \alpha \mathbf{i} + \cos \beta \mathbf{j} + \cos \gamma \mathbf{k}. \tag{3-8}$$

Then, by using Equation 3-5 and Equation 3-6, we have

$$\mathbf{C} = \frac{\mathbf{F}}{\|\mathbf{F}\|} = \frac{F_x}{\|\mathbf{F}\|} \mathbf{i} + \frac{F_x}{\|\mathbf{F}\|} \mathbf{j} + \frac{F_x}{\|\mathbf{F}\|} \mathbf{k}. \tag{3-9}$$

Equation 3-9 can be rearranged for the vector $\mathbf{F}$ in terms of the direction cosines. That is,

$$\mathbf{F} = \|\mathbf{F}\| \mathbf{C}$$

$$= F \cos \alpha \mathbf{i} + F \cos \beta \mathbf{j} + F \cos \gamma \mathbf{k} \tag{3-10}$$

Equation 3-10 implies that if the direction cosines are known, an expression for the vector may be obtained.

EXAMPLE 3.2:

Solve Example 3.1 by using direction cosines.

SOLUTION

In this case, there are only two direction cosines that must be found. These are related to angles α and β, which the *tail* of the vector makes with the x- and y-axes, respectively. The angles are shown in Figure 3.5. Their values may be readily obtained from geometry. In fact,

$$\alpha = 90° - 30° = 60°,$$

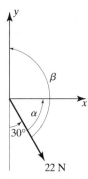

Figure 3.5.

and

$$\beta = \alpha + 90° = 60° + 90° = 150°.$$

Then, by using Equation 3-5 and Equation 3-6, the components are obtained as

$$F_x = F \, \text{Cos} \, \alpha = 22 \, \text{N} \cdot \text{Cos} \, 60° = 11 \, \text{N} \rightarrow$$

$$F_y = F \, \text{Cos} \, \beta = 22 \, \text{N} \cdot \text{Cos} \, 150° = 11\sqrt{3} \, \text{N}\downarrow$$

These are the same results obtained in Example 3.1.

PRACTICE!

A car is climbing a hill that has a slope of 20°, as shown in Figure 3.6. The car weighs 1,800 lb. The weight of the car may be assumed to act as shown in the figure. What are the components of this force in the x and y directions for the coordinate system shown in the figure?

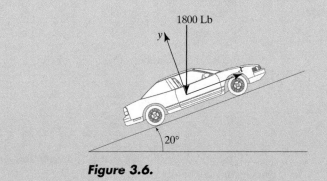

Figure 3.6.

3.2.3 Addition of Vectors

Because many forces may act on a rigid body and since forces are vectors, to solve problems involving vectors, one must be able to find the resultant of the applied forces by adding the vectors. This may be done by using a geometric technique or by summing the components of the force vector.

Vectors may be added geometrically "by joining the vectors head to tail" or "tip to tail" and making use of geometric trigonometric relations. When using this approach, use the following sequence of steps:

1. Construct a sketch by rearranging the vectors so that the head of one vector touches the tail of the other. Be sure to maintain the directions of the vectors.

2. Draw a vector from the tail of the first vector to the head of the last vector. This is the resultant vector.

3. Use geometry or trigonometry to obtain the magnitude and direction of the resultant vector.

If the vectors are drawn to scale, then the direction and the magnitude of the resultant vector may be obtained graphically through direct measurement on the drawing. If direct measurement is not possible, then the unknown quantities may be obtained by using the sine or the cosine law from geometry. If the vectors form a 90° triangle, then trigonometry may be used.

The geometric method works well for two coplanar vectors, but for systems with more than two vectors, extensive calculations are required. In such cases, it is better to add the components of the vector. Of course, this requires that the vectors first be written in terms of their components, which is the reason for discussing the material in Section 3.2.1 and Section 3.2.2. When using the component summation approach, use the following steps:

1. Obtain the components of each vector. This may be done by finding the projection of the vector along the desired direction, as discussed in Section 3.2.1, or by finding the direction cosines and using Equation 3-10 as discussed in Section 3.2.2.

2. Calculate the components of the resultant vector by adding the components of the vectors along the appropriate direction.

EXAMPLE 3.3:

The magnitude of a vector is the sum of its components. Show that the solutions of Example 3.1 and Example 3.2 are correct by adding the components.

SOLUTION
We add the components by placing the two vectors head to tail as shown in Figure 3.7. Let **A** be the sum of these two vectors, which makes an angle α with respect to the horizontal.

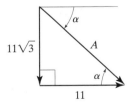

Figure 3.7.

The magnitude of **A**, $\|\mathbf{A}\|$, is obtained by using the Pythagorean theorem. That is,

$$\|\mathbf{A}\| = \sqrt{(11)^2 + (11\sqrt{3})^2} = 22,$$

which is the magnitude of the initial vector.

The angle this vector makes with the horizontal may be obtained from geometry. Since

$$\tan \alpha = \frac{11\sqrt{3}}{11},$$

it follows that

$$\alpha = \tan^{-1}\left(\frac{11\sqrt{3}}{11}\right) = 60°.$$

This means that the vector makes an angle of 30° with the vertical, as defined in the original problem statement to Example 3.1.

PRACTICE!

For the two vectors shown in Figure 3.8, find the resultant vector and the angle it makes with the x-axis.

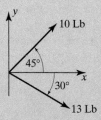

Figure 3.8.

3.3 FORCES, COUPLES, AND MOMENTS

According to Newton's first law, a force is any action that causes a body to move, stop, or accelerate. A force may be represented as a vector.

Now suppose that two parallel forces, F, are applied to a rigid body as shown in Figure 3.9. The forces are a distance d apart and are coplanar; that is, they lie in the same plane. Furthermore, the forces have the same magnitude, but act in opposite directions. The action of such a system of forces causes the body to rotate. The forces may be replaced with a moment, M_c, which is equal to the magnitude of the forces times the distance between them. Mathematically,

$$M_c = Fd. \tag{3-11}$$

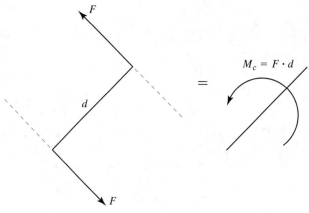

Figure 3.9.

Equation 3-11 gives the magnitude of a couple. The direction of the couple is given by the right-hand rule, with the thumb indicating the direction of the moment and the figures in the direction of the forces. Therefore, a positive couple generates rotation in a counterclockwise sense.

Couples are free vectors; that is, they may be applied at any point P on a body and added in a manner similar as that used to add vectors.

A moment is a measure of rotation about an axis due to a force. The magnitude of a moment is given by Equation 3-11. Again, the right-hand rule is used to indicate a positive moment.

EXAMPLE 3.4:

A man is pushing a 3-foot-high dresser across a room. If he applies a horizontal force of 100 lbf, find the equivalent moment of the force that would cause rotation about the point o in Figure 3.10.

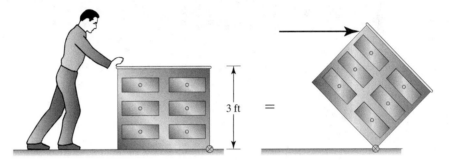

Figure 3.10.

SOLUTION

The equivalent couple is equal to the force times the moment arm. The moment arm is the perpendicular distance between point o and the direction of the force. This moment arm is equal to 3 ft. Thus, the moment is

$$M = F \cdot d = 100 lbf \cdot 3ft = 300 lbf \cdot ft.$$

This moment acts in a clockwise fashion about an axis through o.

PRACTICE!

Water exits from the arms of a sprinkler with a force equal to 2 N, as shown in Figure 3.11. This causes the sprinkler to spin. What is the equivalent moment (in N·m) acting on the sprinkler? Does the sprinkler spin clockwise or counterclockwise?

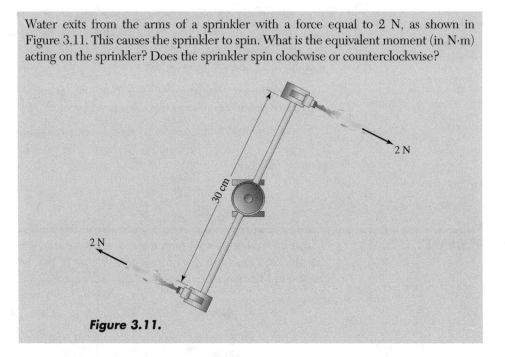

Figure 3.11.

3.4   EQUILIBRIUM AND FREE-BODY DIAGRAMS

A rigid body is in static equilibrium if the resultant force acting on the body is zero. For a system in static equilibrium, Newton's second law implies that the acceleration is zero. Notice this does not mean that the velocity of the body is zero; rather, it means that the velocity is a constant. If **F** and **M** are forces and moments on a system, then for a system in static equilibrium, Newton's second law may be written as follows:

$$\Sigma \mathbf{F} = 0,$$

$$\Sigma \mathbf{M} = 0.$$

(3-12)

The symbol Σ is used to indicate a sum. In particular, $\Sigma \mathbf{F}$ indicates the sum of all the forces acting on the rigid body.

Because Equation 3-12 is in terms of vectors, the equations are often written in terms of the components of the forces and moments. The solution of Equation 3-12 for different types of loading forms the crux of statics. To solve Equation 3-12, the forces and moments acting on a body must be represented within the context of what is happening physically to the body. This is accomplished by constructing a free-body diagram.

A free-body diagram is a sketch showing the loads acting on a rigid body. These loads may be forces and moments. The loads may be internally or externally applied. In solving problems involving rigid-body mechanics, it is imperative that an accurate free-body diagram be constructed.

Free-body diagrams represent the physical state of the rigid body. That is, the loads on the rigid body must be accurately represented so that the physical state of the body is represented.

Students often have difficulty constructing a proper free-body diagram. Because free-body diagrams are so important in statics and dynamics, such students will have difficulty with the courses.

When constructing free-body diagrams, use the following steps:

1. Isolate the body from its physical surroundings, and draw a simplified representation of the body.

2. Pick an appropriate coordinate system, and sketch on your drawing.

3. Add the appropriate loads to the sketch, bearing in mind what is happening physically to the body. Do not forget to add internal loads if needed.

4. Label known quantities with the magnitude and direction.

3.5   FRICTIONAL FORCES

In problems involving the contact of two bodies, if the contact is perfectly smooth, then there will only be a reaction normal to the contact surfaces. On the other hand, if the contact is not smooth, a reaction will occur along the line of contact. This reaction is a force of resistance called the friction. Frictional forces inhibit or prevent slipping.

Frictional forces occur in fluids and solid bodies. In fluids, friction occurs between a fluid and a solid surface, such as between air and a turbine blade. Friction also occurs between different phases of fluids. In solid bodies, friction occurs at the contact surface and is called *coulomb*, or *dry*, friction.

Provided that there is no slipping at the contact surface and that the body is not accelerating, experimental studies have shown that the frictional force is related to the normal contact force by the equation

$$f_s = \mu_s N, \tag{3-13}$$

where f_s is the static frictional force and N is the normal contact force. The constant μ_s is called the coefficient of static friction. Values for μ_s have been experimentally measured for various bodies in contact. However, the value varies due to several factors, including surface finish, humidity, and temperature. For example, for the contact between metal and ice, the value of coefficient of static friction has been obtained in the range 0.03 to 0.05.

If the body is accelerating, but not slipping, then the frictional force has a value less than the static value. This frictional force, f_s, is called the kinetic frictional force and is related to the normal force as

$$f_k = \mu_k N, \tag{3-14}$$

where μ_k is the coefficient of kinetic friction. Values of μ_k are as much as 25% smaller than values for μ_s.

In cases where slipping is occurring at the contact surface, then the frictional force may be less than the value given by Equation 3-13 and Equation 3-14. In solving problems involving slippage, the normal force must be obtained independently of the frictional force.

EXAMPLE 3.5:   What force is required to begin pushing the 15-kg crate shown in Figure 3.12 up the side of a hill? The hill has a 30° slope, and the coefficient of static friction is 0.30.

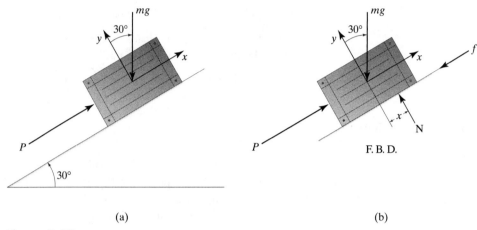

(a)   (b)

Figure 3.12.

SOLUTION

Let the unknown force be P. Because of the position of the force, it is possible that the crate will tip, rather than slide.

Whether the crate will tip depends on the position of the normal load. Let us call this position x, as shown in the figure. We will assume that the crate is on the verge of sliding and calculate the value of x. If x is less than 0.3 m, then our assumption is correct.

If we assume sliding, then

$$f = \mu_s N = 0.3N. \tag{1}$$

Now, we apply Newton's law, Equation 3-12, along each coordinate direction. Summing the forces in the x direction gives

$$\xrightarrow{+} \Sigma F_x = ma_x \Rightarrow P - mg\sin 30 - f = 0. \tag{2}$$

Summing the forces in the y direction gives

$$+\uparrow \Sigma F_y = ma_y \Rightarrow N - mg\cos 30° = 0. \tag{3}$$

Taking the sum of the moments about o gives:

$$+ \ \Sigma M_o = ma_y \Rightarrow -f(0.1) + N(x) = 0. \tag{4}$$

Substituting Equation (1) into Equation (4) and solving for x, we obtain

$$-f(0.1) + N(x) = -0.3N(0.1) + N(x)$$

$$\Rightarrow x = 0.03.$$

Because $0.03 < 0.3$, the crate is sliding, not tipping, and the frictional force is related to the normal force by Equation 3-13, as we assumed.

The value of the normal force is obtained from (3). Thus,

$$N = mg \cos 30° = 15 \text{ kg} \cdot 9.81 \frac{\text{m}}{\text{s}^2} \cos 30° = 127.4 \text{ N}.$$

The value of the force P may be obtained from (2), that is,

$$P = mg \sin 30 + f = 15 \text{ kg } ?9.81 \frac{\text{m}}{\text{s}^2} \sin 30 + 0.3?(127.4 \text{ N})$$

$$P = 111.8 \text{N}$$

EXAMPLE 3.6:

Suppose that the crate from Example 3.5 is not moving. What is the value of the frictional force to prevent the crate from sliding?

SOLUTION

Because the crate is not moving, the frictional force is not given by Equation 3-13. Applying Newton's Laws to the free body diagram shown in Figure 3.13, we obtain

$$+\blacklozenge F_y = ma_y ? N - mg \cos 30° = 0$$

for the y direction and for the x direction:

$$??^+ \blacklozenge F_x = ma_x ? f - mg \sin 30 = 0$$

These two equations may be solved for the normal force and the frictional force. Thus,

$$N = mg \cos 30° = 15 \text{ kg } ?9.81 \frac{\text{m}}{\text{s}^2} \cos 30° = 127.4 \text{ N}$$

and

$$f = mg \sin 30 = 15 \text{ kg } ?9.81 \frac{\text{m}}{\text{s}^2} \sin 30 = 73.6 ?\text{N}$$

Notice that the frictional force is not equal to the value given by Equation 3-13, that is, $(0.3) N = (0.3)(127.4) = 38.2 N.$

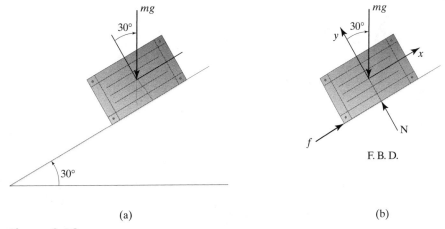

(a)

(b)

Figure 3.13.

3.6   MOTION OF A RIGID BODY

If the loads on a body are unbalanced, then the body will accelerate according to Newton's second law. The study of the acceleration of rigid bodies is undertaken in dynamics. Under the action of unbalanced forces and moments, Equation 3-12 becomes

$$\Sigma \mathbf{F} = m\mathbf{a},$$
$$\Sigma \mathbf{M} = I\alpha,$$

(3-15)

where m is the mass of the rigid body, $\mathbf{a}$ (in m/s^2) is the acceleration of the body's center of mass, I is called the mass moment of inertia (in kg·m^2 or slug·ft^2), and α is the angular acceleration of the center of mass. The angular acceleration is analogous to linear acceleration and has units of rad/s^2.

The branch of rigid-body mechanics dealing with dynamic loads is divided into two fields of study. The first of these fields, kinematics, deals with determining quantities such as position, velocity, and acceleration. The second field of study, kinetics, deals with determining forces and moments.

Acceleration $\mathbf{a}$, is defined as the rate of change of velocity. The average acceleration, $\mathbf{a}_{avg}$, is the change in velocity divided by the change in the time interval.

PROFESSIONAL SUCCESS

When constructing a free-body diagram for an accelerating body, the procedure is as follows:

1. Isolate the body from its physical surroundings, and draw a simplified representation of the body.

2. Pick an appropriate coordinate system, and sketch it on your drawing.

3. Add the appropriate loads to the sketch, bearing in mind what is happening physically to the body. Do not forget to add internal loads if needed.

4. Label known quantities with the magnitude and direction.

However, the acceleration and angular acceleration must be indicated on the diagram. Place these terms at the center of the mass of the rigid body.

EXAMPLE 3.7:   A crate (see Figure 3.14a) accelerates as it slides down an incline, due to its weight. The incline is 30° to the horizontal. Draw the free-body diagram for the crate. You may neglect the air resistance, but do not neglect friction between the crate and the incline.

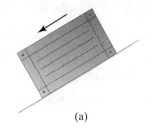

(a)

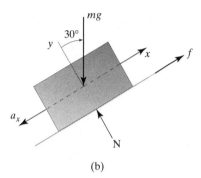

(b)

Figure 3.14.

SOLUTION

The forces on the crate include a frictional force, which acts in the direction opposite to the direction of motion of the crate. Let the force, which acts normal to the incline surface, be indicated by the letter N. By placing a coordinate system as shown in Figure 3.14b, we see that there is no motion in the y direction and hence no acceleration in the y direction.

The free-body diagram is shown in Figure 3.14b. Notice that the acceleration is placed so that it acts through the center of mass.

EXAMPLE 3.8:   Assuming that the coefficient of friction is equal to 0.3, calculate the acceleration of the crate from Example 3.7.

SOLUTION

To calculate the acceleration, we apply Equation 3-15 along each coordinate direction. Thus, the sum of the forces in the y direction gives

$$+\!\uparrow\Sigma F_y = ma_y \Rightarrow N - mg\cos 30° = 0. \tag{1}$$

Therefore, the normal force has a value of

$$N = mg\cos 30. \tag{2}$$

Summing the forces in the x direction gives

$$\xrightarrow{\;+\;} \Sigma F_x = ma_x \Rightarrow -mg\sin 30 + f = ma_x. \tag{3}$$

Assuming sliding without slip, we can relate the frictional force, f, directly to the normal force N. Thus, we have

$$f = \mu N = 0.3N = 0.3(mg\cos 30°), \tag{4}$$

where we have used Equation (2). Equation (4) may be substituted into Equation (3). This gives a relationship for the acceleration:

$$a_x = -g\sin 30° + (0.3)g\cos 30°$$

$$= g(0.3\cos 30° - \sin 30°).$$

Using $g = 9.81$ m/s² and the values of $\cos 30°$ and $\sin 30°$ gives

$$a_x = 9.81\,\frac{m}{s^2}(0.3\cos 30 - \sin 30) \tag{5}$$

$$a_x = 2.36\,\frac{m}{s} \leftarrow .$$

PRACTICE!

A 30-lb, 1-ft-diameter wheel is rolling along the ground due to a 30-lb · ft couple moment. The moment of inertia of this wheel, I, is 0.750 slug · ft². Draw the free-body diagram for this wheel. What is the linear and angular acceleration of the wheel if the static and kinetic coefficients of friction between the wheel and the ground are 0.3 and 0.25, respectively?

3.7   SUMMARY

Rigid-body mechanics, which includes statics and dynamics, is a branch of science that deals with forces and motion of bodies that do not deform under the applied loads. Rigid-body mechanics is one of the fundamental building blocks upon which subsequent courses in mechanical engineering are developed. Thus, the importance in understanding the material in this subject cannot be over emphasized.

A diagram called a free-body diagram may be established for a physical system. In such a diagram, the body under consideration is isolated from its surroundings, and loads acting on the body are shown. The direction and magnitudes of the loads must be properly indicated or the analysis will fail.

Writing a force as a vector, which has magnitude, as well as a direction, allows one to solve problems in rigid-body mechanics involving forces. The projection of a vector along a coordinate direction is called a component of the vector. The components of the vectors may be obtained through geometry. Vectors may be added or subtracted by using a geometric approach. If the vectors are known in terms of their components, then by adding the corresponding component from each vector, the vectors may be summed.

KEY TERMS

components of a vector	direction cosines	free-body diagram
couple	equilibrium force	Newton's laws vector

3.8 EXERCISES

E1 Find the resultant of the vectors shown in Figure 3.15 and the angle the resultant makes with the x-axis.

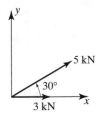

Figure 3.15.

E2. Find the resultant vector, as well as the angle that the vectors make with the x-axis, for the vectors shown in Figure 3.16.

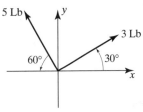

Figure 3.16.

E3. Considering Figure 3.17, find the value of the resultant force, **F**, so that the forces are in equilibrium.

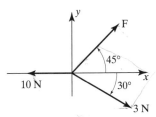

Figure 3.17.

E4. For the system of forces shown in Figure 3.18, find the value of the resultant force, **F**, so that the system is in equilibrium.

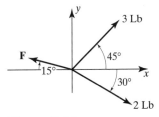

Figure 3.18.

E5.   Find $\mathbf{F}_1$ and $\mathbf{F}_2$ shown in Figure 3.19, so that the system of forces is in equilibrium.

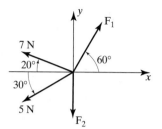

Figure 3.19.

E6.   A 10-foot ladder is leaning against the side of a building, making a 30° angle with the ground, as shown in Figure 3.20. Suppose that the weight of the ladder is taken at its midpoint, draw the free-body diagram of the ladder.

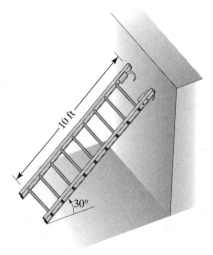

Figure 3.20.

E7.   A rocket is launched vertically as illustrated in Figure 3.21. Draw the free-body diagram of the rocket.

Figure 3.21.

E8.   A cannon ball is shot out of a cannon as shown in Figure 3.22. Draw the free-body diagram of the ball if air resistance is neglected.

Figure 3.22.

E9.   An aircraft is in steady flight when the forces due to drag (D), lift (L), thrust (T) and the weight of the aircraft are in equilibrium, as shown in Figure 3.23. Suppose the weight of the aircraft is 20,000 lb and the two engines each provide 50,000 pounds of thrust. Calculate the values of the lift and drag forces.

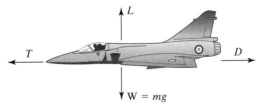

Figure 3.23.

E10. The spring and block shown in Figure 3.24 model the suspension of a certain automobile. Suppose the block has an equivalent mass of 3-kg. A force of 30 N is applied to press the mass downward, compressing the spring as shown in the figure. The spring resists this action with an equal an opposite force, what is the magnitude of this force? Do not neglect the weight of the block.

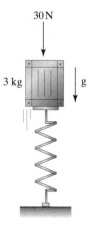

Figure 3.24.

E11. The car in Figure 3.25 is accelerating up along a 20° incline. What force **F** must be applied so that the 1.5-ton car achieves an acceleration of 15 ft/s² in the x direction?

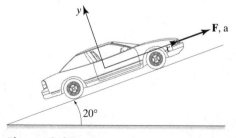

Figure 3.25.

4

Mechanical Engineering and Solid Mechanics

4.1 INTRODUCTION

During the analysis of an engineering design, a mechanical engineer is often faced with predicting the deformation of a body. The starting point for this analysis is statics. Reactions are calculated in response to applied loads, and principles of solid mechanics are used to predict how much the body will deform.

In some cases, the inverse problem is solved. That is, the maximum amount of desired deformation is known and the load that will produce the deformation is desired.

To perform a structural analysis, a relationship between the applied loads and the deformation is required. In practice, large deformations are to be avoided, because they may lead to an unsafe design. Hooke, as we will see in the sections that follow, assumed a linear relationship between the applied loads and the deformation. This relationship is now known as Hooke's law.

4.2 TENSION, COMPRESSION, SHEAR, AND TORSION

If we take a metal bar with a uniform cross section and pull on both ends of the bar, the bar will increase in length, as shown in Figure 4.1. The bar is said to be in a state of tension. The amount that the bar will stretch is determined by the atomic characteristics of the material from which the bar is made.

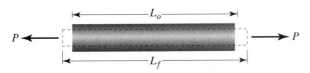

Figure 4.1.

OBJECTIVES

After reading this chapter, you should be able to do the following:

- Differentiate between the different types of basic loading conditions.
- Apply Hooke's law to basic loading conditions.
- Understand how extension of a bar is associated with lateral contraction of the same bar.

Suppose that the bar has an initial length L_0 and it increases to a value L_f. Then, the difference between these two lengths is the change in length $\Delta L = L_f - L_0$.

The engineering strain, a nondimensional parameter, is defined as the ratio of the change in length to the original length. In this case, because the extension of the bar occurs along the longitudinal axis of the bar, the strain may be also termed the axial strain or longitudinal strain.

The strain, ε, may be written as

$$\varepsilon = \frac{\Delta L}{L_0}. \tag{4-1}$$

The stress is the load acting on a unit area. If the bar is pulled apart by a uniform load P and the cross-sectional area of the bar A is uniform, then the stress will be uniform throughout the bar. Because this stress acts normal to the cross section, it is often called the normal stress. The stress may be written as

$$\sigma = \frac{P}{A}. \tag{4-2}$$

Now suppose that the direction on the bar is reversed, instead of pulling on the ends of the bar, we push the ends together. This is compression. Ideally, the bar will decrease in length by an amount ΔL, while the cross section remains constant. Because the bar decreases in length, the strain will have a negative sign.

EXAMPLE 4.1:

A 10-lb sign is suspended from a ceiling with a 1-in-diameter steel rod. What is the stress in the rod if the entire weight of the sign acts through the rod?

SOLUTION
The rod is in tension because of the weight of the sign. The stress is given by Equation 4-2. If d is the diameter of the rod, then the cross-sectional area of the rod is $\pi d^2/4$. Then, according to Equation 4-2,

$$\sigma = \frac{P}{A} = \frac{10\ \text{lb}}{\pi 1^2 in^2/4} = 12.7\ \text{psi}.$$

PRACTICE!

What is the final length of the rod if it has an initial length of 5 in and a strain of 6×10^{-6}?

Suppose we take a straight bar, restrain the bar from moving laterally, and place a load perpendicular to the longitudinal axis as shown in Figure 4.2.

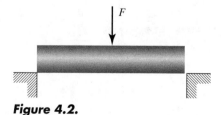

Figure 4.2.

In this case, the bar will be in a state of shear under the action of the shear force F_s. Internally, a shear stress, τ, is developed in response to the shear force that acts over an area A. This shear stress is perpendicular to the longitudinal axis of the bar and may be written as

$$\tau = \frac{F_s}{A}. \tag{4-3}$$

A bar in compression and tension will deform, so that the length of the bar is changed. But how does the bar deform under shear?

Suppose we cut a rectangular element from the bar, as shown in Figure 4.3, where the loads acting on the element are the shear stresses. We assume that the element is stress free in the direction perpendicular to the page.

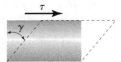

Figure 4.3.

Under the action of the shear stresses, distortion occurs, so that the right angles of the element are changed to some value γ, where γ is called the shear strain.

We may take our uniform bar and clamp one end. Placing two loads P acting in opposite directions and eccentric to the longitudinal axis as shown in Figure 4.4 will create a torque at one end of the bar. Note that if d is the perpendicular distance between the loads, then a torque T equal to P times d will twist the bar, and the bar is said to be in torsion.

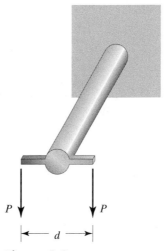

Figure 4.4.

The effect of torsion is to create an angular displacement of one end of the bar with respect to the other. Furthermore, because the loads P are placed laterally to the bar, shearing loads are created at any cross section along the length of the bar.

For a bar of circular cross section, the relationship between the shear stress and torque is

$$\tau = \frac{Tr}{J},$$ (4-4)

where r is the distance from the longitudinal axis to the desired point on the cross section of the bar and J is the polar moment of inertia. The polar moment of inertia is in terms of the geometry of the cross section. For a solid bar of diameter D, the polar moment of inertia is

$$J = \frac{\pi D^4}{32}.$$ (4-5)

PRACTICE!

What is the maximum shear stress in a 4-in circular rod if a torque of 35,000 lb · in is applied? (*Hint*: the maximum shear stress occurs at $r = 2$ in.)

PROFESSIONAL SUCCESS

In general, more than one type of stress may be active in a solid body, due to combined loading conditions. When faced with an engineering problem, an engineer must recognize if more than state of stress exists. In many cases, tension, shear, and so on, may be combined to obtain the correct state of stress in the body. However, because stresses are vector quantities, care must be taken when adding the terms together.

EXAMPLE 4.2:

A load is applied to a half-inch bolted connection as illustrated in Figure 4.5. What is the average shear stress in the bolt?

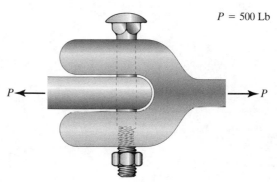

Figure 4.5.

SOLUTION

The average shear stress may be calculated by assuming that half of the applied load acts on each face of the connection. Then the shear force at each face of the bolt is $F_s = P/2 = 2500$ lb. If the bolt has a circular cross section, then the area that the shear force acts over is

$$A = \pi \frac{d^2}{4} = \pi \frac{(0.5)^2}{4} = \frac{\pi}{16} \text{ in}^2.$$ (1)

Then by substituting, this into Equation 4-3, we get

$$\tau_{ave} = \frac{F_g}{A} = \frac{2500 \text{ lb}}{\pi} \cdot 16 \text{ in}^2 = 12732 \text{ psi}.$$

4.3 THE POISSON EFFECT

When a tensile load is applied to a uniform bar, the increase in the length of the bar is accompanied by a decrease in the lateral dimension of the bar. Ideally, the same behavior occurs for a bar in compression, except that there is an increase in the lateral dimension corresponding to a decrease in the length of the bar.

The decrease or increase in the lateral dimension is due to a lateral strain, which is proportional to the strain along the axial direction. The ratio of the lateral strain to the axial strain is related to the Poisson ratio, named after the mathematician who calculated the ratio by molecular theory. Many textbooks use the symbol ν to indicate the Poisson ratio. The Poisson ratio is defined as

$$\nu = -\frac{\text{lateral strain}}{\text{axial strain}}. \tag{4-6}$$

The minus sign in Equation 4-6 is needed in order to keep track of the sign in the strain. For example, because tension corresponds to a decrease in the lateral direction, the lateral strain is negative.

Values of the Poisson ratio have been measured and are available in material handbooks. Typically, the Poisson ratio is in the range 0.25 to 0.35.

PRACTICE!

Suppose that a 2-inch-diameter bar is extended under a tensile load. An axial strain of 3×10^{-6} is obtained. What is the average diameter of the bar after the load is applied?

4.4 HOOKE'S LAW

In 1678, Robert Hooke conducted a series of experiments whereby he measured the extension of metal wires by hanging a mass at the end of the wires. In doing so, Hooke became the first to experimentally investigate the properties of materials. Hooke reasoned that in the deformation of elastic solids, a linear proportionality exists between the applied load and the deformation. Today, we express Hooke's observation in terms of a law named after him. This law is formulated in terms of the stress and strain and may be written as

$$\sigma = E\varepsilon, \tag{4-7}$$

where E is a material constant known as Young's modulus.

Hooke's law may be extended to three-dimensional states of stress under combined loads.

EXAMPLE 4.3:

Suppose that a 4-inch-diameter round bar is extended with a 50,000-lb axial load. The bar has an initial length of 5 feet and extends 0.006 inches. What is the Young's modulus for the material from which the bar is made?

SOLUTION

We can obtain the Young's modulus by using Hooke's law, and Equation 4-1 and Equation 4-2. The stress in the bar is

$$\sigma = \frac{P}{A} = \frac{P}{pd^2/4} = \frac{50{,}000 \text{ lb}}{p4^2} \cdot 4 = 3979 \text{ psi}.$$

The strain is

$$\varepsilon = \frac{DL}{L_0} = \frac{.006 \text{ in}}{5 \text{ ft} \cdot 12 \text{ in}} \text{ ft} = 0.0001.$$

Thus, the Young's modulus for this material is

$$E = \frac{\sigma}{\varepsilon} = \frac{3979 \text{ psi}}{0.0001} = 39.79 \times 10^6 \text{ psi}.$$

PROFESSIONAL SUCCESS

The simple loading cases considered in this chapter form the basics of the study of strength of materials. Structural engineers have used the simple techniques illustrated in the preceding sections to solve a multitude of problems. For complicated machines and structures, this often involved employing assumptions so that simplified models from the theory of strength of materials could be used.

In the 1940s, aircraft engineers working on increasingly complicated aircraft structures initiated the development of the finite-element method. This method is a numerical method characterized by an approach where the machine or structures are divided into small regions. Within each region, the theory of strength of materials is applied, including, if appropriate, Hooke's law. Each region is related to the next region by maintaining continuity on the boundaries. Today, with the advent of powerful desktop computers, the finite element is used to solve problems involving complicated geometries or loading conditions. In fact, it is hard to find an industry involved in structural analysis that does not employ the finite-element method.

4.5 SUMMARY

A solid material under applied load will deform. This deformation may be under tension, compression, shear, and torsion, depending on how the loads are applied. It is also possible that one or more of these loading cases may exist simultaneously.

In the realm of engineering design, it is important to know the behavior of a structure under applied loads. Unsafe designs are characterized by extensive deformations. In such cases, the engineer must redesign the component or structure. The theory of solid mechanics is used to predict the amount of deformation under applied loads and hence whether or not a design is safe.

Hooke assumed a linear relationship between applied structural loads and deformation. This relation is now known as Hooke's law.

KEY TERMS

compression	Poisson ratio	tension
Hooke's law	shear	torsion

4.6 EXERCISES

E1. A 10,000-lb load, as shown in Figure 4.6, loads a structural bar with a square cross section. If the cross section is 2 inches on each side, what is the stress in the structural member?

Figure 4.6.

E2. A concrete support for a bridge is loaded in compression. The cross section for the pedestal is shown in Figure 4.7. What is the value of the average compressive stress if the compressive load is 30 MN?

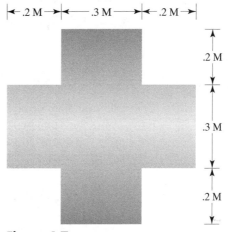

Figure 4.7.

E3. A structural member having two different cross-sectional areas is loaded in tension as shown in Figure 4.8. The applied load is 10 kips. What is the stress in each section?

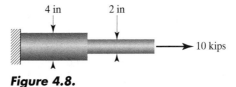

Figure 4.8.

E4. A square hollow tube used in an aircraft structure is loaded in tension by a 25-kip force. The cross section is shown in Figure 4.9. What is the normal stress in the tube?

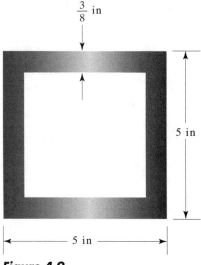

$\frac{3}{8}$ in

5 in

5 in

Figure 4.9.

E5. The structural member shown in Figure 4.10 is used to tie a second-story floor into the structure of a building. Suppose the floor is applying a 2000-lb load on the support, as shown in the figure. What is the average shear in the support? The support extends 6 inches from the plane of the paper.

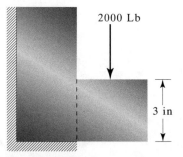

2000 Lb

3 in

Figure 4.10.

E6. As shown in Figure 4.11, 210-MPa load is applied to a piece of wood, which has been cemented to two other boards. The composite board is to be used as a laminated beam in home construction. Each board is a 2 × 4 (actual dimensions 1.5 × 3.5 inches) and 2 feet long. What is the average shear stress in the cemented joints?

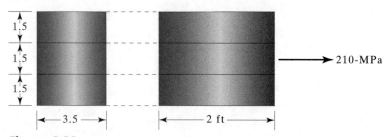

1.5

1.5

1.5

210-MPa

← 3.5 →

← 2 ft →

Figure 4.11.

E7.   An axle for a stamping machine is constructed by bolting together two 3-inch-diameter shafts as shown in Figure 4.12 with three 20-mm bolts. The bolts lie on a 100-mm-bolt circle. A 7-kN·m torque is applied at each end of the shafts. What is the average shear stress in each bolt?

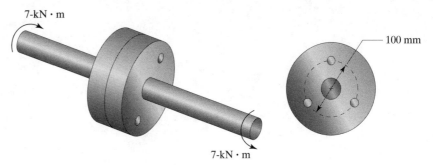

7-kN · m

100 mm

7-kN · m

Figure 4.12.

E8.   Figure 4.13 illustrates a 150-mm circular post embedded in concrete. If a 25-kN·m torque is applied to the rod, what is the shear stress at the rod–concrete interface?

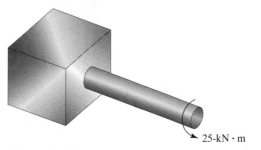

25-kN · m

Figure 4.13.

E9.   A two-inch circular bar is loaded in tension. A strain of 0.0002 is measured. If the bar is constructed of a material with a 0.35 Poisson ratio, find the corresponding lateral strain.

E10. For Problem E1, calculate the corresponding strain if the Young's modulus is 25×10^6 psi.

E11. For Problem E2, calculate the corresponding strain if the Young's modulus is 110 Gpa.

E12. For Problem E4, calculate the corresponding strain if the Young's modulus is 50,000 ksi.

5

Materials and Mechanical Engineering

5.1   INTRODUCTION

Materials are used in the construction of engineering products. Therefore, an engineer must understand which materials are appropriate for which design. The mechanical engineer must have a firm understanding of the properties of the material, which characterize the material's behavior under applied loads and environmental conditions.

Some of these properties are mechanical properties, which relate loading to deformation. Other properties have to do with the behavior of the material in an environment. Examples of such properties are temperature, corrosion, and electrical stability.

5.2   MECHANICAL PROPERTIES OF MATERIALS

The mechanical properties used in engineering are determined by performing a tensile test. Figure 5.1 shows a small "dog bone" specimen placed between two grips. Typical test machines may test the specimen in different ways including tension and compression.

To ensure uniformity in the test, the American Society for Testing and Materials (ASTM) has standardized the test procedure. This standard takes into account the size of the specimen and the gage length, the distance between two gage marks in which the deformation of the specimen is measured. In standardized tests, the gage length is 2.0 inches. The standard diameter of an ASTM specimen is 0.5 inch.

OBJECTIVES

After this chapter, you should be able to do the following:

- Understand how mechanical properties are used to qualify materials for engineering design.
- Discuss how mechanical properties are obtained from a tensile test.
- Understand how traditional and composite materials are used in engineering design.

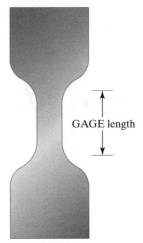

GAGE length

Figure 5.1.

In a static tensile test, the gage length is measured using an extensometer, and a load–extension, or stress–strain, curve is produced. Figure 5.2 shows a typical stress–strain curve for a metal. Characteristics of the curve include a linear region and a region of rapid elongation known as the plastic region.

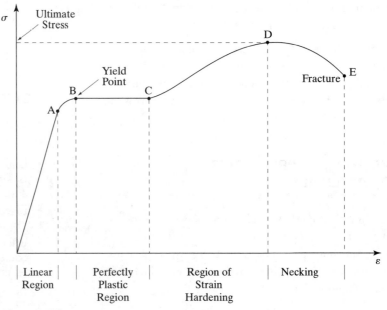

Figure 5.2.

The point at which the linear region ends (point A in the figure) is called the proportional limit. For most metals, this limit occurs at 30 to 100 ksi (200 to 700 Mpa), depending on the carbon content in the steel. Mild steels tend to have a lower limit at about 30 to 40 ksi, whereas high strength steels, with a large carbon content, have a proportional limit of 80 to 100 ksi. The slope of the curve in the linear region is called the Young's modulus. In engineering formulas, this constant is often denoted with the symbol E.

The proportional limit, at point B in Figure 5.2, defines the point in the stress–strain curve where a small increase in the stress yields a large deformation. This phenomenon is called yielding, and the corresponding stress at point B is called the yield point.

Because of this large increase in strain during yielding, most engineering designs tend to avoid the plastic region. The plastic region may have a very small slope. In fact, the curve may be flat. In such a case, the curve is said to be perfectly plastic, and deformation occurs without an increase in the applied load. This is shown in Figure 5.2 in the region from point B to point C.

At some point C in the curve, the steel begins to strain harden; that is, the material undergoes changes in its atomic and crystalline structure. These changes result in an increase resistance to deformation.

Because of the Poisson effect, an increase in the length of the specimen corresponds to a decrease in the lateral dimension of the specimen. At point D, the load reaches the ultimate stress, and the reduction of the cross section of the specimen is substantial. A small neck develops in the specimen. This process is called necking. The cross section in the neck is much smaller than the rest of the specimen, and hence the local stress is much higher. Reduction in size of the neck continues until the cross section is too small to support the applied load. At this point (point E in Figure 5.1), the specimen fractures.

The actual shape of the stress–strain curve varies from material to material. Some materials undergo large deformations before failure and are called ductile materials. Lead, mild steels, nickel, aluminum, and brass are a few examples of ductile materials. Ductile materials are often characterized without a clearly defined yield point.

On the other hand, some materials such as glass and ceramics exhibit little or no ductile behavior. These materials exhibit a linear or nearly linear relationship between stress and strain from the inception of loading until fracture.

Mechanical properties obtained from such tests as the one just described are tabulated in mechanical handbooks. Perhaps foremost of these properties is the value for Young's modulus. For aluminum and aluminum alloys, the Young's modulus is about 10.5 ksi (73 Gpa). For steels, the constant has a value of 28–30 ksi (190–210 Gpa).

If the goal of a design is to design within the linear region, then all stresses in the structure or component must be below the yield stress. Because there is a range in the yield stress for most materials, engineers apply a safety factor to their design. This safety factor can be as much as two to three times the actual load in the structure or component.

Metals and many other traditional engineering materials are assumed to be homogeneous and isotropic. A *homogeneous* material is a material for which the elastic properties are the same at all points in the elastic body. If the elastic properties are the same in all directions at any point on the body, the material is said to be an *isotropic* material.

Composite materials, which are constructed by combining two or more constituent materials at the macroscopic level, are nonisotropic and often nonhomogeneous.

PRACTICE!

Use reference material from your school library to find the proportional limit, yield point, ultimate strength, and Young's modulus for cast iron.

PROFESSIONAL SUCCESS

Notice that the mechanical properties discussed in this section are obtained through a tensile test. Also, the results are for statically applied loads. For cases involving combined loadings, a failure criterion is used.

Failure theories are beyond the scope of this book. However, it is interesting to note that various failure theories have been proposed for cases involving combined loadings or advanced materials.

5.3   MATERIALS AND THEIR USE IN ENGINEERING DESIGN

Material science is concerned with the creation of new materials and the application and processing of materials by understanding the composition and structure of the material.

Properties of a material are defined by the internal characteristics of the material. The atomic structure and the bonds between the atoms influence internal structure of materials. Flaws in the material, such as cracks and voids, also influence properties of materials. Such influences are the focus of active research by material scientists, as well as engineers.

To properly apply a material, a mechanical engineer needs to have an understanding of how the internal structure influences the behavior of the material. In doing this, the engineer may consider mechanical properties such as Young's modulus, Poisson ratio, and yield stress, as well as temperature and corrosion. For example, plastics have found wide applications in engineering design, because they are lightweight and strong. However, a plastic material would be an unsuitable material from which to construct a furnace, because it would melt!

Heat treatments affect the properties of some materials, including metals. Heat treatments can be designed and utilized to achieve materials with superior qualities. For example, swords have been tempered since archaic times by dipping the heated metal in a pool of water. This procedure is called quenching. Quenching causes the precipitation of a second phase in the metal. This second phase influences the material behavior at the microscopic level by blocking the propagation of defects in the crystalline structure called *dislocations*.

The density of a material often becomes a criterion for material selection. Aluminum, although not as strong as steel, is less dense and therefore for a given volume would weigh less. Aluminum alloys are used extensively in constructing aircraft components, since every pound saved readily translates into additional revenue for the airline. Recently, to minimize fuel consumption, automobile manufacturers have shifted to aluminum alloys and polymer materials for structural components, where steel had been used previously.

Besides considering a material's properties with respect to the design goal, an engineer must also consider how the properties will influence the manufacturing of the design. For example, materials that are very hard will require tooling made from even harder materials; otherwise, the tooling will wear out in a very short period. Tools made of hard materials often translate to increased tooling and hence manufacturing costs. This will increase the cost of the design.

In general, an engineer must consider the properties of a material to properly choose the material for the design, while bearing in mind the difficulty in processing the materials. This may seem like a formidable task, and often it is, but it is an aspect that makes engineering challenging.

EXAMPLE 5.1:   A gurney to be used in a medical evacuation helicopter is being designed. A strong lightweight material is desired. The design calls for design within the linear range with a yield stress, based on the service load, of about 65 ksi. What is a candidate material?

SOLUTION
Aluminum alloys are among the most lightweight materials, are readily available, and can be easily worked. Typical aluminum alloys have specific weight weight per unit volume of 160 to 180 lb/ft^3 compared with a value of 490 lb/ft^3 for steel. Thus, an aluminum

alloy would be an excellent candidate from the point of view of weight. The Young's modulus for aluminum alloys ranges from 10,000 to 11,400 ksi, whereas for steel, the range is 28,000 to 30,000 ksi. If the value of E is suitable for the design, then the aluminum alloy is a better candidate, based on its low specific weight. The aluminum alloy 7075-T6, which is used extensively in aircraft construction, has a yield stress of 70 ksi, which would make the 7075-T6 and ideal candidate.

PRACTICE!

Polymers such as ultrahigh-density polyethelyene are used in bioengineering applications as replacements for arteries, veins, and small joints. Use the Internet or reference material in your school library to obtain information regarding the advantages and disadvantages in using this material for such applications.

5.4   THE USE OF ADVANCED MATERIALS: COMPOSITES

A composite material is a material formed by the combination of two or more chemically distinct and insoluble constituent materials at the macroscopic level. In engineering practice, this combination is such that it improves the overall properties of the entire structure, such as weight, strength, and stiffness. Often, the constituents are combined, so that the best properties are available along a given direction.

Composite materials may be grouped into three separate types, as shown in Figure 5.3. These categories are as follows:

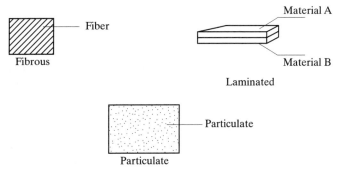

Figure 5.3.

1.   Fibrous composites are constructed by imbedding stiff fibers in a ductile matrix. An example of a fibrous composite material is fiberglass in a polymer matrix. Graphite tennis racquets and fishing poles are constructed with fibrous composites. Such materials have graphite fibers imbedded in a matrix made of epoxy resin.

2.   Laminated composites are constructed by bonding together two or more layers of a different material in sheets to form a single structure. Plywood is a commonly used laminate composite.

3.   Particulate composites are constructed by imbedding metallic or nonmetallic particles in a matrix. Concrete is an example of such a composite, with the cement acting as the matrix and the gravel as the particulate.

The use of composite materials in engineering design is more involved than traditional materials, because the mechanical properties vary from point to point in the material and in different directions. Thus, composites are nonhomogeneous and nonisotropic. A mechanical engineer needs to be aware of this when using composites. Nevertheless, composite materials are finding increased use in manufactured items.

PRACTICE!

Composite materials are used extensively in aerospace applications. Use the Internet or reference material in your school library to find information concerning the use of composite materials in the new Boeing 777 design. What structural components are composed of composite materials? What types of composites are used?

PROFESSIONAL SUCCESS

In the past, the use of composite materials in engineering practice was limited primarily to aerospace and defense systems. In such industries, the motivation for using composite materials was and still is the high strength-to-weight ratio. Recently, spurred by environmental, safety, and economic reasons, we have begun to use composite materials in new areas of engineering practice.

For example, because of the higher strength of the composite, utility poles may be constructed of composite material and used as replacement for poles previously taken from old growth forests. Single poles made of composite materials have the same load-carrying capability of structural towers used to carry power cables over vast rural areas. The poles have a much smaller footprint than the traditional towers. Thus, they may be placed along right of ways where space is restricted.

In addition, power transmission lines have been proposed where aluminum wires are clad with a composite material. Because the conductivity of the composite is close to that of copper, the composite clad aluminum wire is able to carry more current than a conventional power cable. The composite also acts as a thermal insulator, constraining the thermal expansion of the aluminum.

5.5 SUMMARY

Material science is the study of the creation of new materials, processing, and application of materials. The application of a material to a given design is influenced by the properties of the materials. New materials are created to have superior properties. These properties are bulk responses of the microstructure of the material.

Properties of materials may be quantified by certain parameters, such as Young's modulus, yield stress, and so on. In applying materials to a design, the engineer must know what property to use in deciding whether the material is acceptable. In some cases, a material that would be acceptable by considering a mechanical property might be unacceptable because of another property, such as weight. Furthermore, an engineer must keep in mind how the properties of the material influence the manufacturability of the design. Materials that lead to high manufacturing costs should be avoided in favor of materials with more sensible manufacturing costs.

Composite materials are finding increased use in engineering design. Such materials are constructed by combining two constituent materials at the macroscopic level to produce a material with superior properties. The three categories of composite materials are fibrous, laminated, and particulate.

KEY TERMS

composite material
curve
mechanical properties

proportional limit
stress–strain
yield point

Young's modulus

5.6   EXERCISES

E1.   Plot the data given in Table 5-1 and obtain Young's modulus, the proportional limit, and the ultimate stress.

TABLE 5-1 Data for Exercise E1

STRESS (KSI)	STRAIN
10	0.0005
15	0.0020
25	0.0025
33	0.0031
45	0.0039
51	0.0050
57	0.0051
61	0.0056
63	0.0057
65	0.0058
66	0.0059
65	0.0060
64	0.0061
fracture	0.0062

E2.   The data in Table 5-2 are obtained by using a 2.00-gage-length test specimen. Initially, the diameter of the specimen is 0.508 inch, but, at failure, the diameter of the neck is 0.445 inch. The total elongation is 0.0045 inch.

TABLE 5-2 Data for Exercise E2

LOAD (LB)	ELONGATION (IN)
1	0.0005
2	0.0010
3	0.0015
4	0.0020
5	0.0025
6	0.0030
7	0.0035
8	0.0040
9	Fracture

E3.   A tension test is applied on an unknown material. Young's modulus is found to be 10,600 ksi, the yield stress is 40 ksi, and the ultimate stress is 45 ksi. Use this information and a material handbook in your library to determine the unknown material.

E4.   Use a material handbook from your library to fill in the data for the materials in Table 5-3.

TABLE 5-3   Table for Exercise E4

PROBLEM	MATERIAL	YIELD STRESS (KSI)	ULTIMATE STRESS (KSI)	YOUNG'S MODULUS (KSI)
5.5	Bronze			
5.6	Steel			
5.7	Glass			
5.8	Rubber			

E5.   Plot the data shown in Table 5-4, and show that these data are for a material having a perfectly plastic range. Obtain Young's modulus for the elastic range.

TABLE 5-4   Data for Exercise E5

STRESS (KSI)	STRAIN
10	0.0004
15	0.0008
20	0.0010
25	0.0015
30	0.0020
35	0.0023
40	0.0030
45	0.0035
45	0.0040
45	0.0050
45	0.0060
45	0.0070

E6.   List some natural occurring composites, and describe some appropriate applications. Use the Internet or your school library to uncover this information.

E7.   Use your library or the Internet to obtain information concerning the application of composites to electrical applications. List several such applications. What are the type of composites employed in these applications?

E8.   In bioengineering applications, it is necessary to model bone as a structural material. Is bone homogeneous or nonhomogeneous? To complete this exercise, use your library or the Internet to search for this information.

6

Fluids and Mechanical Engineering

6.1   INTRODUCTION

A fluid is a substance that deforms continuously under an applied shear stress. Fluids include liquids, such as water, and gases, such as air. For a Newtonian fluid, proportionality exists between the shear stress and the rate of deformation. For a fluid, the deformation is equal to the velocity gradient (i.e., the rate at which the velocity changes in the fluid). Thus, for a Newtonian fluid, we may write

$$\tau = \mu \frac{\partial u}{\partial y}, \tag{6-1}$$

where u is the velocity of the fluid, y is a coordinate direction, τ is the shear stress, and μ is a constant called the *viscosity* (in $\mathrm{N \cdot s/m^2}$). The viscosity is a property of the fluid, though it does vary with temperature. If the viscosity is zero, then the fluid is said to be *inviscid*. Inviscid fluids do not actually exist. However, there are many problems where the assumption of an inviscid fluid might be made to simplify the analysis and still yield meaningful results.

Sometimes, the viscosity is given in terms of the density. Let ρ be the density of the fluid. Then, the *kinematic viscosity* ν is defined as

$$\nu = \frac{\mu}{\rho}. \tag{6-2}$$

Fluid mechanics is concerned with Newtonian fluids, which are at rest or moving. Because fluids often interact with a structure (e.g., a river pushing against a dam), the determination of forces characterizing the interaction are often the focus of the analysis.

OBJECTIVES

After reading this chapter, you should be able to a do the following:

- Understand the difference between a fluid in motion and a fluid at rest.
- Find the pressure in a hydrostatic fluid.
- Use Newton's law and calculate the force on a surface due to a fluid.

6.2   FLUIDS AT REST

Many engineering applications involve fluids in a state of rest; that is, the fluid is not moving. A branch of fluid mechanics called hydrostatics addresses such problems. An example of a fluid at rest is coffee in a coffee cup. The coffee is a liquid that is not moving, yet it exerts a force on the walls of the cup. The force on each wall is related to the pressure in the fluid.

Anyone who has ever dived into a swimming pool clearly knows that the pressure is highest at the bottom of the pool. This is because the weight of each element of the water is pressing down. At the bottom of the pool, the entire volume of the water is bearing its weight on the diver.

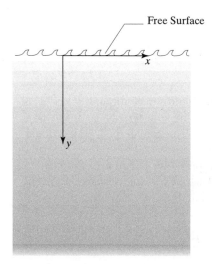

Figure 6.1.

To formulate an expression for this phenomenon, let us take a fluid and place a coordinate system at the top of the fluid with positive y acting in the direction of increasing depth, as shown in Figure 6.1. The pressure increases as y increases and is affected by the weight of the fluid per unit volume. The weight of the fluid is ρg, where ρ is the density of the fluid. Also, suppose that the pressure at the surface of the fluid, the free surface, has some value p_0. Thus, we have

$$p = p_0 + \rho g y. \tag{6-3}$$

The force acting on the bottom of the cup is equal to this pressure times the area of the bottom of the cup.

This equation only works for a surface that is uniformly submerged below a fluid. For surfaces inclined with respect to the free surface, a differential form of Equation 6-3 must be obtained.

Buoyancy, or Archimedes' principle, is a restating of Equation 6-3. This principle states that a submerged body is subject to an upward force F_B equal to the weight of the fluid displaced. The principle may be expressed as:

$$F_B = \rho g \times \text{volume}. \tag{6-4}$$

EXAMPLE 6.1:

What is the pressure at the bottom of a three-meter-deep swimming pool?

SOLUTION

The pressure is given by Equation 6-3, where P_0, the pressure at the free surface, is equal to one atmosphere of pressure (101 kPa). The density of water is 1000 kg/m³. Thus, Equation 6-3 yields

$$p = p_0 + \rho g y,$$

$$p = \left(101 \times 10^3 \, pa\right) + 1000 \, \frac{kg}{m^3} \cdot 9.81 \cdot \frac{m}{s^2} \cdot 3 \, m,$$

$$p = 101.81 \text{ kPa.} \qquad 131 \, K \, Pa$$

PRACTICE!

A hot-air balloon rises off the ground because the density of the air in the ballon is less than the density of the atmosphere. Suppose that a spherical balloon has a diameter of 10 feet and that the density of the air is 0.071 lbm/ft³. How much weight can the balloon lift and remain aloft? For this analysis, neglect the cooling of the air in the balloon.

PROFESSIONAL SUCCESS

Hydrostatic pressures variations are neglected in hydraulic systems, because they are quite small compared with the pressure of the fluid. For example, auto-mobile hydraulic brakes can develop pressures as high as 10 MPa (1500 psi) and aircraft and machinery hydraulic systems have pressures up to 30 MPa (4500 psi).

6.3 FLUIDS IN MOTION

For a fluid at rest, the velocity is zero. However, for a fluid in motion, the velocity becomes an unknown quantity and may vary from point to point in the fluid. At points where the fluid interacts with another surface, the fluid will have the same velocity as the surface. This is called the no-slip condition. For example, the velocity at the bottom of the river is zero, because the water is in contact with the ground, which is not moving.

In solving problems involving fluids in motion, two approaches may be taken. The first of these is to track every particle of the fluid as it moves. However, because fluids deform continuously, this would require tracking the particle continuously. Another approach is to pick a region in the fluid and watch the fluid as it enters and leaves the region. This second approach is called the *control volume* approach and is more practical for solving engineering problems involving the flow of a fluid. In using the control volume approach to solve problems, conservation laws such as conservation of mass and the laws of Newton are applied.

EXAMPLE 6.2:

Suppose we are designing a water-piping system for a building with two apartments and wish to supply water to two faucets, which are connected by a tee configuration, as shown in Figure 6.2. We want the velocity of the water to be same when it leaves both faucets. What should be the velocity of the water source?

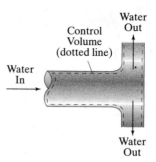

Figure 6.2.

SOLUTION

We pick a control volume, as shown in Figure 6.2, and apply mass conservation. In this problem, it is practical to assume that the mass of the fluid per unit time entering the control volume is equal to the amount leaving. This will be true if the source of the water remains constant and if fluid is not trapped within the control volume. Let us label the inlet as region 1, with a corresponding velocity v_1, and the two outlets as regions 2 and 3. The mass per unit time is called the mass flowrate and is equal to the density times velocity and the cross-sectional area of the pipe. Then, conservation of mass gives

$$\dot{m}_{in} = \dot{m}_{out},$$

$$\dot{m}_1 = \dot{m}_2 + \dot{m}_3,$$

$$\rho v_1 A_1 = \rho v_2 A_2 + \rho v_3 A_2,$$

where A_1, A_2, and A_3 are the cross-sectional areas of the pipe at Section 6.1, Section 6.2, and Section 6.3, respectively. If the density of the water is the same throughout, then the density term ρ will drop out of every term in the equation. Furthermore, if the diameter of the pipe is the same everywhere, then the cross-sectional areas are the same everywhere. Then, we have

$$v_1 = v_2 + v_3.$$

However, we want $v_1 = v_2$. Thus,

$$v_1 = 2v_2 = 2v_3.$$

Therefore, in order for the velocity to be the same at each outlet, v_1 must be twice the value at Section 6.1 or Section 6.2.

In Example 6.2, we assumed that the water was steady (i.e., that the water was not trapped in control volume). Had water accumulated in the control volume, the amount of water leaving would not be equal to the amount entering the control volume.

A fluid with constant density is called an incompressible fluid. Compressible fluid flow comes into consideration in fluids that are moving at high speed, such as the area around an aircraft.

If the speed of a fluid exceeds the speed of sound, the flow is called supersonic. Qualitatively, new phenomena, such as shock waves, occur at supersonic speeds. The phenomenon of shock waves are very complex and beyond the scope of this book.

Nevertheless, it is important to note that shock waves need to be taken into consideration when designing supersonic aircraft and projectiles.

Moving fluids pose some practical engineering problems that have only recently been understood. For example, aircraft flight only became possible in the last century because investigators did not fully understand the mechanics of fluid flow. It is the interaction of air and the wing of an aircraft that allows the aircraft to become airborne.

The section of an aircraft wing is called an airfoil. The upper and lower surfaces are curved. The upper surface has more curvature than the lower, as shown in Figure 6.3. Because the upper surface is more curved, the air must speed up in order to flow over the surface. For a moving fluid, an increase in velocity corresponds to a decrease in pressure. Thus, a smaller pressure exists on the top surface of the wing than on lower surface. The higher pressure essentially pushes the wing up.

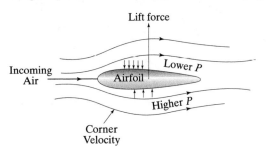

Figure 6.3.

Engineers called aerodynamicists can calculate this force, which results due to this imbalance in pressure. This force is called the lift.

It is hard for a fluid particle to flow at high speed on a very curved path. Thus, if the airfoil is too thick or if the angle that the airfoil makes with the horizontal is too large, the fluid will not be able to remain attached to the upper surface of the airfoil. As shown in Figure 6.4, if the fluid ceases to move across the top of the wing, then the wing is stalled, and the imbalance in the pressure no longer exists. Stall may lead to a catastrophic failure of the aircraft.

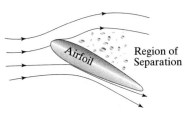

Figure 6.4.

As previously stated, if the momentum of a fluid is known, then forces may be obtained by using Newton's second law. In general, formulating an expression suitable for solving engineering problems requires the use of calculus. However, if we assume steady one-dimensional flow, then we may write that the force in the kth direction, F_k, as

$$F_k = \dot{m}(V_{2k} - V_{1k}), \tag{6-5}$$

where $\dot{m}$ is the mass flowrate and $(V_{2k} - V_{1k})$ is the difference in the velocity of the fluid from station 1 to station 2 in the kth direction.

EXAMPLE 6.3:   Water with a velocity of 10 m/s strikes a turbine used for power generation and is rotated 60° from the horizontal by the blade, as shown in Figure 6.5. The cross section of the inlet water is 0.003 m². What is the force on a turbine blade shown in the figure?

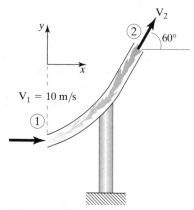

Figure 6.5.

SOLUTION

In solving this problem, we employ Equation 6-5 and the concept of the conservation of mass. We may assume incompressible flow, so that $\rho_1 = \rho_2$. For this problem, we also assume that $A_1 = A_2$. The fact that the water is initially horizontal means that $V_{1y} = 0$.

Applying the idea of conservation of mass indicates that

$$\dot{m}_1 = \dot{m}_2 = \rho_1 V_1 A_1 = \rho_2 V_2 A_2 = \dot{m}$$

$$\dot{m} = 1000 \, \frac{kg}{m^3} \cdot 10 \, \frac{m}{s} \cdot 0.003 \, m^2$$

$$\dot{m} = 30 \, \frac{kg}{s}.$$

Notice that because $A_1 = A_2$, then $V_1 = V_2$.

Applying Equation 6-5 to the x direction yields

$$F_x = \dot{m}(V_{2x} - V_{1x}) = \dot{m}(V_2 \cos 60° - V_1) = \dot{m}V_1(\cos 60° - 1).$$

Substituting the appropriate numbers into this expression gives

$$F_x = 30 \, \frac{kg}{s} \cdot 10 \, \frac{m}{s} \cdot (\cos 60 - 1) = 150 \, N \leftarrow .$$

The arrow indicates the direction of the x component of the force.

Likewise, in the y direction, we have

$$F_y = \dot{m}\left(V_{2y} - V_{1y}\right) = \dot{m}\left(V_2 \sin 60° - 0\right) = \dot{m} V_1 \sin 60°$$

$$F_y = 30 \frac{\text{kg}}{\text{s}} \cdot 10 \frac{\text{m}}{\text{s}} \sin 60° = 259.8 \text{ N}\uparrow.$$

PRACTICE!

What would be the angle of the turbine in Example 6.3 if the vertical force $\left(F_y\right)$ was zero? What would be the value of the horizontal force $\left(F_x\right)$?

PROFESSIONAL SUCCESS

To understand how a fluid reacts around complicated machinery, fluid mechanists use experimental methods and visualization techniques. One method, which works well with air, involves injecting a stream of smoke into the air stream. A model of the object, such as an aircraft, is placed in a wind tunnel and the smoke injected. As the smoke flows past the aircraft, it creates a series of lines called streaklines. Each streakline represents all the fluid particles passing through that point in the fluid.

6.4   SUMMARY

In engineering problems involving fluid mechanics, the fluid may be moving or may be at rest. If the fluid is at rest, then the velocity of the fluid is everywhere zero. Solution of such problems requires obtaining the value of the pressure in the fluid. The pressure will linearly increase from the free surface.

On the other hand, if a fluid is moving, the velocity of the fluid becomes a variable and may vary from point to point in the fluid. One approach to solving engineering problems with moving fluids is to pick a region of the fluid called a control volume and apply conservation laws such as conservation of mass and momentum to the volume. Newton's second law may be used to obtain forces, because this law relates force to the change in momentum.

Control volumes must be properly chosen. In problems involving the interaction of a fluid with a structure, the control volume is chosen so as to include the structure or the reaction of the structure to the fluid.

KEY TERMS

control volume
density

hydrostatic pressure

viscosity

6.5   EXERCISES

For problems E1–E4, use a handbook to find the viscosity for each fluid at the given temperature.

E1.   SAE 30 oil at 20° C.

E2.   SAE oil at 100° C.

E3.   Kerosine at 20° C.

E4.   Kerosine at 100° C.

E5. What is the pressure and force acting on the rectangular underwater gate shown in Figure 6.6? The gate is 1-m wide by 1-m deep.

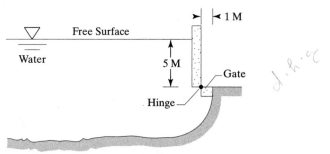

Figure 6.6.

(handwritten notes in margin:) $P_t = P_o + \rho h g$

$1 \, atm + 1000 \cdot 5 \cdot 9.81$

$\frac{kg}{m^3} \quad m \quad \frac{m}{s^2}$

$161,300 \, P$

E6. A tank, as illustrated in Figure 6.7, contains two layers of water, each of which is at a different temperature and hence a different density. The densities of each fluid are such that $\rho_1 = 1000$ kg/m³ and $\rho_2 = 1500$ kg/m³. What is the hydrostatic pressure at the bottom of the tank?

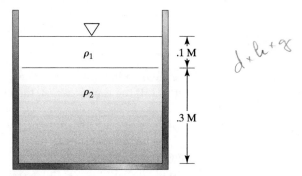

Figure 6.7.

(handwritten note in margin:) $d \times h + g$

E7. In Figure 6.8, water enters a converging two-dimensional channel with a velocity of 5 m/s. If the inlet is 0.5-m wide, the exit is 0.2-m, and the channel is 0.5-m wide perpendicular to the plane of the paper, what is the velocity of the fluid at the exit?

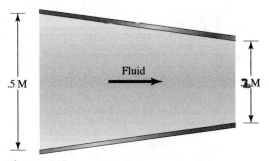

Figure 6.8.

E8.   Air flows through a ventilation duct. The section is a rectangular box shown in Figure 6.9. If the fluid enters at stations 1 and 2 and leaves at station 3, what is the velocity of the fluid at station 3? The areas at the stations are $A_1 = 0.1$ m², $A_2 = 0.3$ m², and $A_3 = 0.15$ m². The velocity at stations 1 and 2 are known to be $V_1 = 1$ m/s and $V_2 = 5$ m/s, respectively.

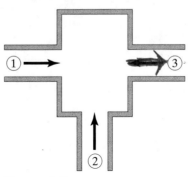

Figure 6.9.

E9.   In Figure 6.10, water strikes a hemispherical cup and is rotated 45°. If the inlet velocity of the water is 5 m/s, what is the force on the cup? The jet of water has an initial diameter of 100 mm and 50 mm at each outlet.

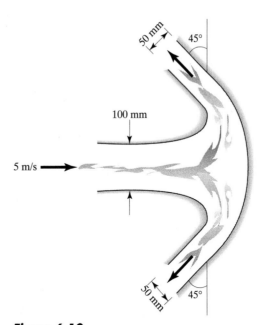

Figure 6.10.

E10. What would be the force on the cup if the cup from Exercise E9 had a 50-mm central hole, as shown in Figure 6.11?

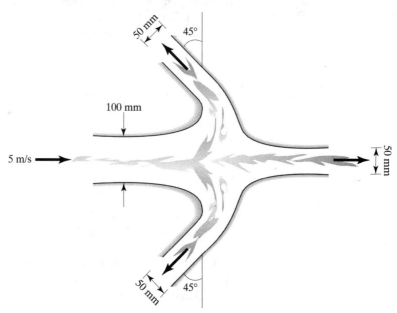

Figure 6.11.

7

The Thermal Sciences and Mechanical Engineering

7.1 INTRODUCTION

Thermal science is an area of scientific thought encompassing thermodynamics and heat transfer. Thermodynamics is a science that deals with heat and work and the properties of substances related to heat and work. Heat transfer is a science that deals with the transfer of thermal energy.

Mechanical engineers are often involved in the design and analysis of machinery wherein the working fluid or substance undergoes a thermal cycle. In undergoing a thermal cycle, the substance passes through a succession of thermodynamic states. Some systems produce work (as in power systems), while others require work input to produce other effects, as in refrigeration.

Examples of such systems include internal combustion engines, refrigerators, heating and air-conditioning systems, and fossil-fuel power plants. The basic objective of these power systems is to take a fuel source and transform the chemical energy contained within that fuel source into work. Often, the net work output is used to propel a vehicle or to produce electric power.

Engineers define the efficiency η of a power system as the net work output divided by the net heat input (i.e., the useful effect divided by the required input), or

$$\eta = \frac{\text{net work out}}{\text{net heat in}}. \qquad (7\text{-}1)$$

The efficiency is often given as a percent. This is obtained by multiplying the ratio from Equation 7-1 by 100.

OBJECTIVES

After reading this chapter, you should be able to do the following:

- Understand the concept of temperature and convert a temperature from one scale to another.
- Discuss the different modes of heat transfer.
- Understand that the design or analysis of a machine operating in a thermal cycle is governed by the first and second laws of thermodynamics.

A power system's efficiency increases as the net work output increases or as the net heat input decreases. Thus, a highly efficient system will produce very much work output, while requiring little heat input.

Air-conditioning systems remove heat from the air in an enclosed environment such as a house. To do this, a refrigerant must be compressed, which requires work input in the form of electrical energy.

In a way similar to that of efficiency, a coefficient of performance (COP) is defined for a refrigeration system as the heat removed by the working fluid from the refrigerated space divided by the required work input:

$$\text{COP} = \frac{\text{heat removed}}{\text{net work in}}.$$

In this case, the useful effect is the heat removed.

In the past 20 years, the cost of fossil fuels has risen considerably because petroleum and natural gas prices have risen. Inefficient systems use more fuel than efficient systems to produce the same effect; therefore, an objective of a mechanical engineer working in designing machinery is to increase the efficiency. To accomplish this goal, the engineer must have a clear understanding of the science involved in the system, as well as the mechanical limitations of the machinery.

7.2 THE CONCEPTS OF TEMPERATURE AND HEAT TRANSFER

What is temperature? We may say it is 10 below zero outside, but what does that mean?

Temperature is a measurement of the energy of molecules due to their rapid movement. Temperature is measured indirectly using a transducer. A typical transducer, the thermometer, measures the temperature by the expansion and contraction of a liquid (usually mercury) in a glass tube. In the case of a mercury thermometer, if the temperature rises, the mercury expands and hence rises in the tube. If the temperature drops, the opposite occurs.

Various scales have been defined to quantify this phenomenon. These scales are somewhat arbitrary. For example, the Fahrenheit scale, named after Daniel Fahrenheit (1686–1736), places the freezing point of water at 32 degrees and the boiling point of water at 212 degrees. The Celsius scale, named after Anders Celsius (1701–1744), places the freezing point at zero degrees and the boiling point at 100 degrees. These values may be used to obtain the following relationship between the Celsius scale and the Fahrenheit scale:

$$°F = 1.8(°C) + 32°; \tag{7-2\,a}$$

$$°C = \frac{5}{9}(°F - 32°). \tag{7-2\,b}$$

The Kelvin scale, named after Physicist Lord Kelvin, is an absolute scale, with zero Kelvin defined as when the motion of all molecules ceases. This point occurs at $-273\,°C$; thus, there is a difference of 273 degrees between the Celsius and Kelvin scales. In particular,

$$K = °C + 273. \tag{7-3}$$

EXAMPLE 7.1:   What is the temperature in °C and Kelvin for 70 °F?

SOLUTION

Equation 7-2 and Equation 7-3 give the conversion to each temperature scale. In Celsius, using Equation 7-2, we see that the temperature would be

$$70°F = 1.8(°C) + 32°,$$

$$°C = \frac{70 - 32}{1.8} = 21 °C.$$

In Kelvin, using Equation 7-3, we see that the temperature would have the value

$$K = 21°C + 273 = 294K.$$

PROFESSIONAL SUCCESS

Notice that when dealing with temperature differences, there is no need to convert from degrees Celsius to Kelvin. This is because a change in one degree in the Celsius scale is equivalent to a change of one unit in the Kelvin scale.

Heat transfer is a thermodynamic process. Heat transfer will occur whenever there is a temperature difference. Heat transfer is quantified in terms of the heat Q (in Joules, J), heat rate $\dot{Q}$ (in Watts, W), or heat per unit area $q''(\text{W/m}^2)$, called the heat flux.

There are three different modes of heat transfer, as shown in Figure 7.1. These modes are conduction, convection, and radiation.

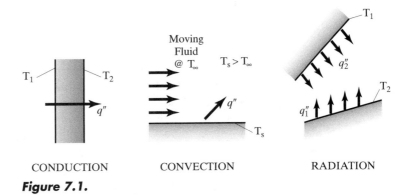

CONDUCTION   CONVECTION   RADIATION

Figure 7.1.

In conduction, the heat is transferred between warmer and colder regions of the same body. In this case, the heat flux is from the high temperature to the low temperature and is caused by the temperature gradient within the body. A law by Fourier relates the heat flux to the temperature gradient as

$$q'' = -k\nabla T,$$

where T is the temperature, k is a proportionality constant called the thermal conductivity, and ∇ is the del operator, or gradient. The term ∇T describes how the temperature varies as a function of position.

The thermal conductivity has units of W/(M·K) in SI and is a property of a material. Insulating materials will have a small thermal conductivity.

If the thermal conductivity is constant throughout the body and the temperature varies with respect to one dimension only, say, x, then Fourier's law can be reduced to

$$q'' = -k\frac{\Delta T}{\Delta x}, \tag{7-4}$$

where ΔT is the change in the temperature and Δx is the change in the x dimension.

In convection, the heat is transferred from a surface to a moving fluid when a temperature difference exists between the surface and the fluid. Heat will be transferred more readily if the fluid is moving rapidly across the surface. An example of this is the wind chill effect. On a cold day, an individual's heat is transferred to the air because a temperature difference exists between the individual and the air. If the wind blows, much more heat is transferred, by convection. The harder the wind blows, the greater the heat transfer. Forced convection is used to describe convection wherein the fluid is propelled by an external means such as a fan, wind or pump.

The heat flux transferred by convection is given by:

$$q'' = k\Delta T \tag{7-5}$$

where h is called the *heat transfer coefficient*. Equation 7-5 is called Newton's law of heat convection.

The heat transfer coefficient has the same units of W/m² in the SI system. The larger the value of the heat transfer coefficient, the larger the amount of heat transfer for a given temperature difference. Typically, the heat transfer coefficient for free convection is in the range $2 \le h \le 25$ W/(m² · K), whereas for a gas under forced convection, the heat transfer coefficient will have a value in the range $25 \le h \le 250$ W/(m² · K). The heat transfer coefficient can be on the order of 10^3 or higher for forced convection involving liquids.

Heat may be transferred between two surfaces without an intervening medium if the surfaces are at different temperatures. This form of heat transfer is called radiation and it propagates in the form of electromagnetic waves. Energy is emitted in radiation by all matter that has a finite temperature because of the changes in the atoms composing the matter. The Stefan-Boltzmann law, Equation 7-6, gives the maximum amount of energy that may be transmitted.

$$q'' = \sigma T_s^4, \tag{7-6}$$

where $\sigma = 5.67 \times 10^{-8}$ W/(m² · K⁴) is a constant called the Stefan–Boltzmann constant and T_s is the temperature of the radiating surface. A surface that radiates energy according to the Stefan–Boltzmann law is called an *ideal radiator*, or *blackbody*. Real surfaces emit radiation at a lower value. An expression for a real radiator may be obtained by multiplying the right-hand side of Equation 7-6 by a factor ε called the emissivity, which has a value between zero and one. Also, suppose that the radiation is occurring to the surroundings, which has a temperature T_{sur}. Then, Equation 7-6 becomes

$$q'' = \varepsilon\sigma(T_s^4 - T_{\text{sur}}^4). \tag{7-7}$$

In many practical applications, heat transfer occurs in one or more of these modes. Often, one or more mode may be predominant, which allows one to neglect the remaining mode(s). This greatly simplifies the model of the heat transfer.

EXAMPLE 7.2:

On a cold day, the temperature outside of a house reaches $-30\,°F$, whereas the temperature inside the house is $70\,°F$. What is the heat transferred through a wall of the house with thickness of 4.5 inches if the wall is made of concrete with a thermal conductivity of 1 W/(m·K)? Assume that heat transfer by radiation and convection is negligible.

SOLUTION
The heat transfer is given by Equation 7-4. First, the temperature values are converted to Celsius. Using the methods shown in Example 7.1, we find that $-30\,°F = -34.4\,°C$ and $70\,°F = 21\,°C$. Also, 4.5 inches = 0.1143 m. Then, by substitution into Equation 7-4, we get

$$q'' = k\Delta T = 1\,\frac{W}{m\cdot °C}\,\frac{(-34.4 - 21)°C}{0.1143m}(-34.4 - 21)°C = 484.7\,\frac{W}{m^2}.$$

PRACTICE!

A wall in a house has a window with the same thickness as the wall. The heat loss through this window is being considered. Suppose that the wall has a thermal conductivity of 0.135 W/(m·K) and that the glass window has a thermal conductivity of 0.60 W/(m·K). Compare the heat flux lost through the window with that through the wall, and suggest a way to reduce the heat loss through the window.

PROFESSIONAL SUCCESS

When solving problems in the thermal sciences, use the following format:

1. Draw a schematic of the system under consideration. Show the boundaries of this system. Identify relevant heat and work transfers.

2. List all assumptions that are going to made in solving the problem.

3. List the property values used in the analysis. Include tables or diagrams where the values were obtained.

4. Apply the laws of thermodynamics in a manner consistent with your assumptions.

5. Consider your results, and draw the appropriate conclusions. Review your assumptions.

7.3 THE FIRST LAW OF THERMODYNAMICS: ENERGY CONSERVATION

A system is a quantity of matter of fixed mass and identity. This system will have a thermodynamic state defined by properties, including temperature, pressure, volume, and mass. This system will undergo a change in thermodynamic state if any of these properties are changed. Energy may be exchanged with the surroundings as work or heat.

This principle of energy conservation, which is the first law of thermodynamics, includes the energy transferred to work and heat. The work may be done by or on the system, while heat may be added to or removed from the system.

In solving engineering problems, a mathematical expression for the first law is needed. This may be obtained by letting E be the change in energy in going from thermodynamic state 1 to thermodynamic state 2, Q the heat transfer in, and W the work done. Then, the first law may be written as

$$\Delta E = Q - W. \tag{7-8}$$

In the special case where two bodies of different temperature are brought in contact, there may be no work done, but heat is transferred. Then, energy conservation requires that the change in energy be equal to the heat transfer, or that

$$\Delta E = Q.$$

The energy of the system E is composed of three separate parts: internal energy U, kinetic energy KE, and potential energy PE. That is,

$$E = U + \text{KE} + \text{PE}. \tag{7-9}$$

Internal energy is defined by the thermodynamic state of the system. A system will have kinetic energy if it is moving. In many applications, the kinetic energy will be

$$\text{KE} = \frac{1}{2}mv^2, \tag{7-10}$$

where m and v are the mass and velocity of the system, respectively.

If the potential energy is due to a reference to a gravitational potential, then the potential energy may be written as

$$\text{PE} = mgz, \tag{7-11}$$

where z is the height above the gravitational reference and g is the gravitational acceleration.

Employing Equation 7-9, Equation 7-10, and Equation 7-11, we find that the first law becomes

$$\Delta U + \frac{m\Delta(V^2)}{2} + mg\Delta z = Q - W, \tag{7-12}$$

where ΔU, $\Delta(V^2/2)$, and Δz are the change in the internal energy, kinetic energy, and height of the system, respectively, from state 1 to state 2.

The work done in compressing or expanding a substance is called pv work, because it is characterized by a change in pressure and volume. The work done by a system during a process between an initial volume v_1 and final volume v_2 is

$$W = \int_{v_1}^{v_2} p\,dv, \tag{7-13}$$

where p_2 and p_1 are the final and initial pressures, respectively, and v_1 and v_2 are the volumes at the initial and final states. Equation 7-13 indicates that the variation of the pressure as a function of the volume v is needed in order to evaluate the work.

EXAMPLE 7.3:　　Air in a 0.01-m³ cylinder is heated as shown in Figure 7.2. The cylinder has a frictionless piston at the top of the cylinder on which several small weights are placed. During the process the volume of the gas doubles. If the initial pressure is 100 kPa, what is the work done in expanding the air?

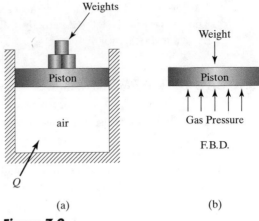

Figure 7.2.

SOLUTION

We recognize that the air undergoes pv work. This is a quasiequilibrium process, wherein deviation from thermodynamic equilibrium is small. Thus, the gas may be considered to be in equilibrium.

As shown in Figure 7.2b, a free-body diagram indicates that the force due to the mass of the weights and the pressure are in equilibrium. Therefore, the pressure of the gas is constant while it expands.

The value of the work is given by Equation 7-13. This work is done on the piston by the gas. By integration of Equation 7-13, we get

$$W = \int_{v_1}^{v_2} p\,dv = p(v_2 - v_1)$$

$$= 100\,\text{kPa}(0.005\,\text{m}^3 - 0.01\,\text{m}^3) = -0.500\,\text{kJ}.$$

EXAMPLE 7.4:

Water flowing over a paddle wheel has been used from ancient times as a source of power. Suppose that water from a river is moving at 3 m/s before falling over a paddle wheel. After striking the paddle, the river is moving at 1 m/s. If the difference in height of the river is 5 m, what is the amount of work per unit mass of the water done by the river in striking the paddle wheel?

SOLUTION

We can assume that the water has the same thermodynamic state before and after striking the paddle wheel. Also, this is an adiabatic process; that is, there is no heat transfer. Then, the only nonzero terms in the first law of thermodynamics [Equation 7-12] are the terms relating to the height and kinetic energy differential. Thus, the work done per unit mass is

$$\frac{W}{m} = \frac{\Delta(V^2)}{2} + g\Delta z = \frac{1}{2}(1^2 - 3^2)\frac{\text{m}^2}{\text{s}^2} + 9.81\frac{\text{m}}{\text{s}^2}(5\text{m}) = 45.1\frac{\text{J}}{\text{kg}}.$$

PRACTICE!

According to the first law of thermodynamics, it is possible for the efficiency of a power system, as defined in the introduction, to have a value of 100 %. Describe how this may be possible.

7.4 THE SECOND LAW OF THERMODYNAMICS

To understand the second law of thermodynamics, let us consider a simple example. Suppose that you have a hot cup of coffee. Because the coffee is warmer than its surroundings, heat is transferred from the coffee to the surroundings according to the first law of thermodynamics. Given enough time, the cup of coffee will reach thermodynamic equilibrium and will not revert to its initial temperature by itself.

The second law of thermodynamics defines the direction of a process. In the case of the cup of coffee, the direction of the process was such that the coffee cooled by heat transfer to the surroundings. Certainly, after the coffee has cooled, it is possible to warm the coffee.

In the case of the cooling of the coffee, heat was transferred from a high temperature to a cooler temperature. If we wanted to transfer heat from a cooler temperature to a warmer temperature, then work input is required. This is how a refrigerator works.

In addition to the direction of a process, the second law of thermodynamics specifies the limits for the efficiency of thermal devices. A device that has a higher efficiency violates the second law and is called a perpetual-motion machine.

PRACTICE!

While it is possible for the efficiency of a power system to be 100% according to the first law of thermodynamics, in practice, the efficiency will be less than this value. Describe how the second law of thermodynamics places an upper bound on the efficiency.

PROFESSIONAL SUCCESS

Mechanical engineers designing a new machine or analyzing existing ones always check for violation of the first and second laws of thermodynamics.

7.5 SUMMARY

Mechanical engineers involved in the design and analysis of machinery must be aware of the change in the thermodynamic state during the cycle of the machinery. Many practical systems take a fuel source and convert the chemical energy of the fuel source to work, heat, or both. In doing so, the total energy must be conserved, as stated by the first law of thermodynamics. Real processes always obey the first law of thermodynamics.

Natural heat transfer is from a body at a high temperature to a body at a low temperature. The potential of heat to go naturally from high to low temperatures can be utilized to produce work through a heat engine. However, not all the available heat can be converted to work, since some heat is always lost to the surroundings.

KEY TERMS

Celsius	Fahrenheit	temperature
energy	heat	laws of thermodynamics

7.6   EXERCISES

E1. What is the temperature of air in Fahrenheit if it has a temperature of 13 °C?

E2. If water is at 70 °F, what is its corresponding temperature in Celsius?

E3. What is the temperature in Kelvin of water if its temperature in Fahrenheit is 50 °F?

E4. What is the temperature in Fahrenheit for air if the air has a temperature of 515 K?

E5. Water in a tank is stirred. If 6000 kJ of work are done to the fluid and 1000 kJ of heat are dissipated, what is the change in the internal energy of the fluid?

Figure 7.3.

E6. Air is contained in a cylinder as shown in Figure 7.3. The piston at the top of the cylinder contains small weights. Initially, the gas is 150 kPa and has a volume of 0.02 m³. Suppose that the weights are slowly removed so that the gas expands according to $pv^{1.2} = $ constant until the volume is 0.1 m³. What is the work done in raising the weights?

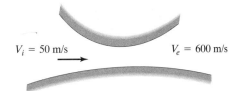

Figure 7.4.

E7. Air enters an insulated nozzle as shown in Figure 7.4 with a velocity of 50 m/s and leaves with a velocity of 600 m/s. What is the change in internal energy of the substance?

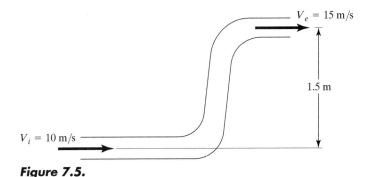

Figure 7.5.

E8.   5000 kJ of work are done to elevate water 1.5 m as shown in Figure 7.5. The process is adiabatic. What is the change in internal energy of the fluid?

E9.   The temperature of the air in a hot-air balloon is much warmer than the surrounding atmosphere. Describe the cooling of the air in the balloon as dictated by the first and second laws of thermodynamics. (See Figure 7.6.) What must be done in order to keep the balloon aloft?

Figure 7.6.

E10.   An automobile engine works by combustion of gasoline. This produces an explosion that drives a piston and in turn a camshaft, as in Figure 7.7. The turning of the shaft returns the piston to its initial position, and the process is repeated. Qualitatively describe how the first and second laws of thermodynamics dictate the heat transfer, the work, and the efficiency of this cycle.

Figure 7.7.

8

Mechanical Engineering and Design

8.1 INTRODUCTION

The design of engineering components forms a large part of the daily activities of a mechanical engineer. Graphical skills are used in at least 90% of the design process. For the rest of the process, verbal, mathematical, and written skills are used. Graphical skills are used most of the time, because it is very difficult to communicate details of a design in written form or verbally.

Because of the extensive use of graphics in design, design becomes a visual process. This ranges from the conceptual phase, where a three-dimensional model of the product is developed, to the refinement phase, where fine detailed drawings of each component are created.

Technology has made graphical communication of a design easier. In CAD, the computer is used to produce graphical images. This is called *geometric modeling*. Graphics created by computer can be easily stored, retrieved, and modified. Before the advent of CAD systems, pencil (or pen) and paper, along with mechanical instruments such as a T-square, scaled ruler, and set of triangles, were used to create the drawings. With such methods, companies were required to employ a large pool of professional draftsmen. While this may have created employment opportunities, it involved large numbers of people, all of whom had to be informed on the latest design changes, making management of a design more difficult.

OBJECTIVES

After reading this chapter, you should be able to do the following:

- Discuss the differences between traditional and concurrent engineering.
- Describe how graphical visualization techniques are used in the design process.

8.2 TRADITIONAL DESIGN VERSUS CONCURRENT ENGINEERING

Considering the work done by engineers, design is the process by which an idea for a commercial product is conceived and communicated. As mentioned in the introduction to this chapter, because the idea must be communicated, the design process becomes a graphical process. CAD systems have greatly enhanced the graphical communication of design ideas, but CAD is merely the method by which ideas are communicated; it is not the process by which the design is performed. Improving the method used in design can directly enhance the development of a design into the finished product.

The traditional approach to design is an iterative one. The design of the product proceeds along a succession of steps, beginning with a conceptual design or ideation phase and ending with a refinement phase.

Figure 8.1 illustrates a traditional approach that may be used to produce a new aircraft design. The design and development of the aircraft is the answer to a request for the aircraft. A specific customer request, a federal request for proposal, or an internal decision may initiate the development of the aircraft. Early in the design process, the design goes through an ideation phase where rough sketches of the design are constructed. These sketches are used to illustrate the pertinent elements of the design. For example, the basic shape, the number of engines, exits, seating arrangements, and so on, are outlined.

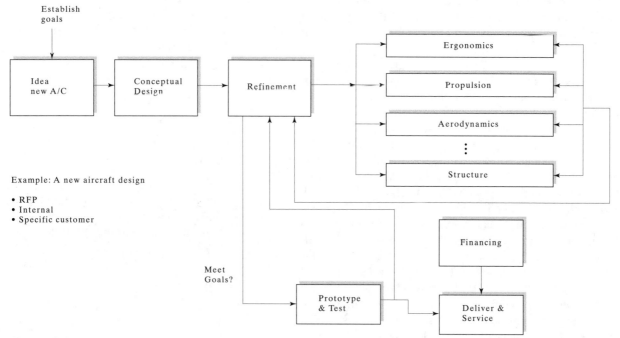

Figure 8.1.

The conceptual design, along with any sketches, form the starting point for the refinement phase of the design. (See Figure 8.1.) In traditional design, the refinement phase is characterized by a number of groups working independently to refine the design. Each group completes a design based on the goals of the group. For example, the aerodynamics group, which is interested in the flight characteristics of the aircraft, creates a design for the flight surfaces of the aircraft, including the wing and tail.

Drawings are used to graphically communicate the design. Because the groups work for the most part independently, a design from one group often conflicts with another. It is the job of the project managers to fine tune the individual group design into one coherent workable solution. From this final design, one or more prototypes may be built and tested. Only after the final design has met all the desired design objectives is the product delivered to the customer. It is at this point that financial elements, such as the sales and marketing professionals, of the organization are brought in to deliver the product to the consumer.

If the aircraft were to be designed using *concurrent engineering* (shown in Figure 8.2), a different approach would be used. While the different design groups are still involved, the organizational structure is such that each group works on a final design that will meet the overall objectives, as opposed to a design to meet the objectives of each group. The advantage is that groups from all areas of the company are involved early in the design process, and because the design groups are closely linked, design conflicts may be solved more readily and efficiently.

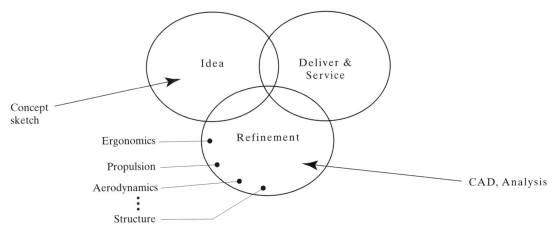

Figure 8.2.

In addition, unlike the traditional approach to design, financial and marketing groups are included early in the design process. Often, the consumer is also involved early on in the design process. In our example of the aircraft design, this would involve, say, United Airlines, early in the design process. Involving the consumer in the design process allows the design to be tailored to the consumer.

The advent of computer technology allows the use of concurrent engineering in engineering design. Thanks to CAD, the current design is always available among the different design groups, financial and marketing elements, as well as production staff. Current computer and CAD technology is such that the design may be developed and analyzed entirely inside the computer, without any need to construct a prototype, which saves money in development costs.

Just as in the traditional approach, design with concurrent engineering involves both a conceptual and a refinement phase. Sketches are used to communicate ideas during the conceptual design phase. Detailed drawings are used in the refinement phase.

PRACTICE!

Describe how the traditional approach to design may be used to develop a high-performance bicycle. What sort of design groups would be involved in the design?

8.3 THE DESIGN PROCESS

Whatever the design approach, two basic phases exist in the design process. This includes the ideation, often called conceptual design, and the refinement phase. The ideation phase is characterized by a period whereby the design progresses from a problem statement to a preliminary design. The refinement phase takes the preliminary design and produces a final design that may be manufactured economically.

The ideation phase includes the following steps:

1. The problem statement, containing a concise statement of the problem to be solved, is formulated.

2. Information gathering is the process through which useful information for the design is collected.

3. A feasibility study is performed. In this step, data are gathered to ascertain whether there is a market for the product. If a competing product exists, that product is examined.

4. The goals of the design are determined. These goals include objectives and limitations of the design.

5. A schedule is developed outlining when various aspects of the design and development will be performed.

6. Using a brainstorming process, multiple solutions are defined. The solutions are weighed with respect to the goals of the project. A preliminary design is established from these solutions. Computer or physical models are created from the preliminary design.

The end result of the ideation phase is a preliminary design. This preliminary design can now be fine tuned to a final design using modeling and analysis. In completing this phase, the rough sketches are taken and used to create detailed dimensionally correct drawings and models. After a material selection process is performed, analysis of the design may begin. This analysis may use finite-element modeling and kinematic analysis or concepts from fundamental mechanics, such as statics, dynamics, solid mechanics, and fluid mechanics principles.

Mechanical engineers are well trained to perform the analysis aspects of a design. However, a mechanical engineer must recognize that while a solution may be technically feasible, it may not be economically, politically, or socially viable. Thus, conflicts may arise between the technical feasibility of a design and the impact of the design on society.

In order for the mechanical engineer to develop skills to deal with these conflicts, many universities require students in engineering to take courses in economics and professional ethics. Furthermore, classes in the humanities and social sciences are required to educate the student, so that when in the workplace, the engineer will have an understanding of the effect any design might have on society.

Certainly, the economics of a design play a significant role in the development of the design solution into a commercial product. After a design is completed, a company must gather the raw materials, pay the production staff, and market the product. In order for the company to remain profitable, these costs must be included in the final cost of the product.

Such added costs may be reduced by careful consideration during the design phase. For example, advanced materials, such as composites, may make a design goal easier to achieve, but may increase the final cost of the product substantially, because of the cost of the raw materials. In this case, the designer must decide whether it is really necessary to use a composite material instead of a traditional material.

In addition, an engineer must keep in mind the effect that the design solution will have on society. Political pressures may come to bear if the design goal goes counter to the political will of society. Often, such pressures are a response to the design's impact on the environment. The supersonic jet airliner Concorde is a prime example. The aircraft was designed for maximum speed over long distances, such as the trip from London, England, to Sydney, Australia. However, the airport authority in Sydney refused landing rights to the airplane, because of concerns regarding the jet's sonic boom and pollution to the atmosphere by its engine exhaust. Many other cities soon followed Sydney's example. For this reason and because of reasons having to do with high fuel costs, the service life of the aircraft has become limited.

The Concorde is an example of a great technical design that met its design goals admirably. Many of the solutions employed to achieve its faster-than-sound speed were unique at the time they were employed. However, the airplane has had limited financial success. In this sense, the design was a failure.

PRACTICE!

A new design for a high-performance mountain bicycle is under consideration. What are some of the design goals that would have to be met? What are some issues that may prevent the design from becoming economically or socially feasible?

PROFESSIONAL SUCCESS

When working on a design, keep in mind the interface with the user. The design should be such that the product is intuitive to use. For example, the VHS videocassette recorder helped to revolutionize how people watch television. However, not long after its introduction into society, the VCR became the butt of jokes, because of the difficulty that users were having in programming the machine to record. Manufacturers of the VCR soon realized this difficulty and developed solutions, including on-screen programming.

8.4   GRAPHICAL TECHNIQUES FOR COMMUNICATING YOUR IDEA

An engineer needs to be able to communicate his or her idea in a clear, concise manner. To accomplish this, a standard graphical approach has been developed using various methods to illustrate three-dimensional objects on a two-dimensional media, such as paper or the computer screen.

Traditionally, pencil (or pen) and paper have been used with descriptive geometry to create standard dimensioned drawings. These drawings usually contain several orthographic views, an example of which is shown in Figure 8.3.

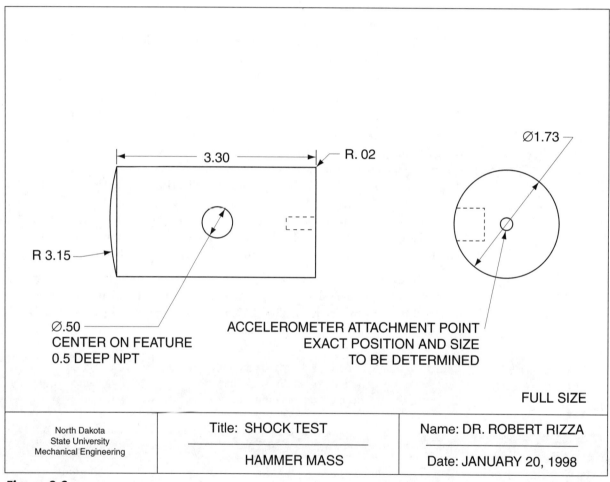

Figure 8.3.

Using an isometric ruled paper, three-dimensional sketches of a model might be created. Isometric paper has diagonal lines at 30° from the horizontal. As shown in Figure 8.4, receding lines in the model are sketched along these diagonals, giving an illusion of three-dimensionality.

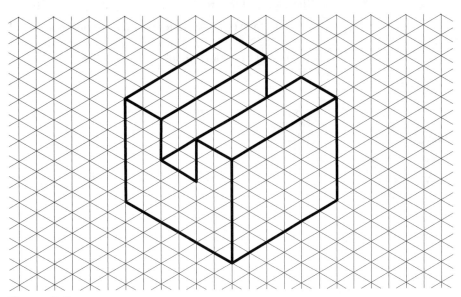

Figure 8.4.

The advent of the computer, and CAD software has made it possible for the creation of graphics images using the computer. The use of CAD allows images to be stored and easily updated.

The latest CAD software packages, such as Pro/Engineer and AutoCad Mechanical Desktop, are parametric feature-based modelers. A parametric modeler, also called a constraint-based modeler, uses parameters to represent a model, as opposed to dimensions. The model is generated using the parameters that may be assigned arbitrary numbers. This allows the model to be updated more quickly and easily, because the numerical values of the parameters need only be updated whenever a change is made.

For example, a model of an automobile cylinder, say, of 2-inch diameter, would be created in the computer using a traditional modeler in terms of its 2-inch diameter. On the other hand, a parametric model would be created in terms of the diameter and the number 2 associated with the diameter.

A feature-based modeler combines frequently used commands to create common manufactured features. For instance, a hole is a common manufactured feature that may be represented on the computer by two circles and two lines connecting the circles. In traditional modelers, each circle and line would have to be individually added by the user. In a feature-based modeler, the software knows how the hole is to be represented; therefore, all the user has to do is input the size of the hole and its location.

Primitive modelers tend to use a finite number of primitive geometric shapes to create the models. These shapes, which are shown in Figure 8.5, are combined to create the model by adding or subtracting the primitive shapes. An example of this process is shown in Figure 8.6.

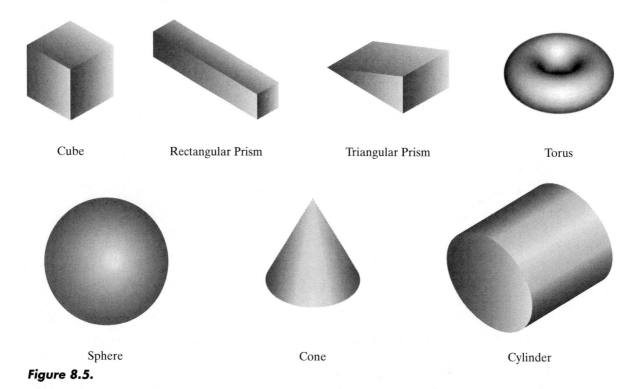

Cube   Rectangular Prism   Triangular Prism   Torus

Sphere   Cone   Cylinder

Figure 8.5.

On the other hand, feature-based modelers use a sketch to outline the basic shape of the feature. (See Figure 8.6.) The sketch does not have to be dimensionally correct, because it is defined in terms of parameters whose numerical values may be changed at any time. After the sketch is constrained, it is extruded, revolved, or swept along a trajectory to create the finished model.

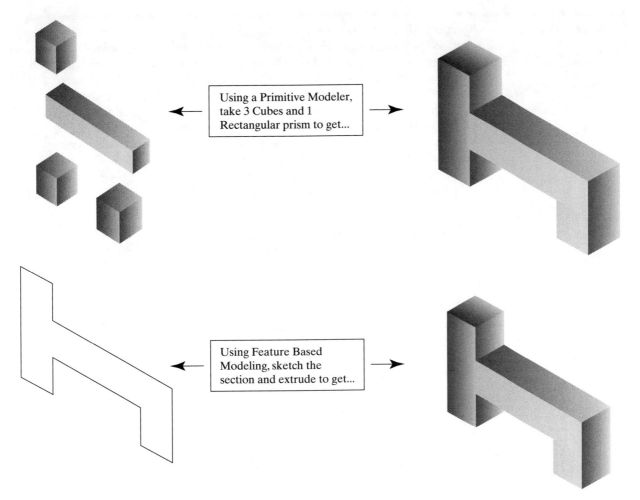

Figure 8.6.

Parametric feature-based modelers are the preferred computer modelers for three-dimensional models. Such modelers, along with concurrent engineering, discussed in Section 8.2, form the modern approach to engineering design.

PRACTICE!

Find out about the CAD package used in your department. Is the modeler a primitive or parametric modeler? How is the software package used in the design sequence within your department?

EXAMPLE 8.1:

Spectrum AEROMED of Wheaton, Minnesota, manufactures medical modules for aircraft and helicopters and was interested in the design of a patient loader that would transfer a patient from ground level to medical evacuation aircraft for the Mayo Clinic. In August 1999, a student group in the mechanical engineering program at NDSU undertook the design as part of their capstone mechanical engineering design class.

The group members spent two semesters to achieve the final design. The design team used the steps outlined in Section 8.3. They began with a statement of the problem, which became the objective of their design. This objective was stated as follows: "to design a lightweight, compact patient loading mechanism that will ensure patient stability during loading into and out of the aircraft."

Spectrum AEROMED supplied specific design criteria. These criteria established goals for the design:

1. Patient must be lifted from the gurney height of 36" (from ground) to the module height of 56" and translated horizontally approximately 33" into the plane.
2. Patient must remain level at all times during loading.
3. System should provide hands-off operation.
4. System must be lightweight (under 45 pounds); leeway (within 10–15 pounds) in the total weight is acceptable.
5. Overall dimensions (when folded for stowing) may not exceed 44 × 16 × 4 inches.
6. The system must have capacity for a 500–pound patient.
7. 500 pounds is the optimum capacity; the current loader (in use) has a capacity of only 350 pounds.
8. Time for full extension/retraction is not to exceed 30 seconds (not including setup time).
9. The loader is to be used with the Mayo Med-Air airplane.

Additional specifications for the design included power- and air-supply information, including a 24 VDC, 115 VAC (from inverters that have 4.3 amps available) electrical supply and a compressed-air supply of 0.42 CFM at 50 psi.

After the design goals were identified, the students began a period of brainstorming wherein several solutions were identified. These designs are illustrated in Figure 8.7–Figure 8.10 and are copies of the original concept sketches. The designs were compared and contrasted with respect to the design goals.

In preliminary design #1, shown in Figure 8.7, a rotating arm drives a bed along a set of curved rails. During the movement, the bed is kept level at all times. However, when this design was analyzed, it was recognized that certain dimensional constraints of the airplane, such as the location of the steps and the patient module, would restrict the movement of the driving arm. Difficulties were also encountered in finding a realistic approach to fold the curved rail into the necessary storage dimensions required by design goal #5.

A parallel linkage was used in design #2, shown in Figure 8.8. The rocker arms allow the loader to use the airplane stairs as a base. As the arms rotate, the bed is driven up and around to the module. The bed slides along rollers in a horizontal fashion at the ends of the rocker arms to the module top. This solution yields a relatively compact design. However, the designers could not find a suitable method to drive the rocker arms. Also, the arms may achieve a "lock-up" position when they are parallel to ground. Another disadvantage to this design is the fact that it is freestanding; that is, it does not attach to any part of the plane. This creates difficulties in setting up the machinery and decreases stability. Design #2 was eliminated, but many of its features were used in the final design.

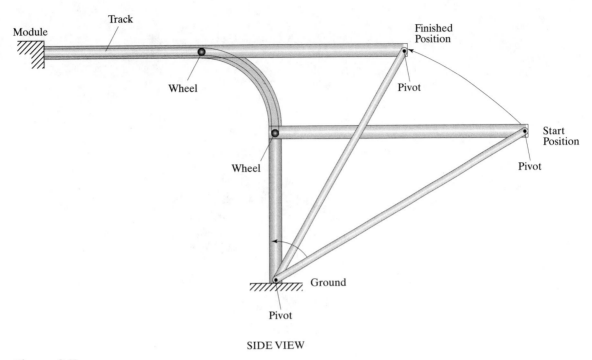

SIDE VIEW

Figure 8.7.

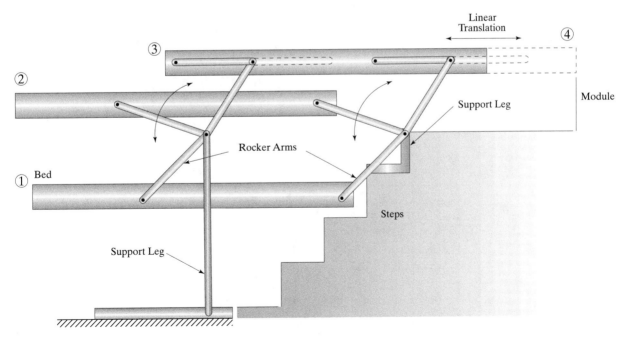

SIDE VIEW

Figure 8.8.

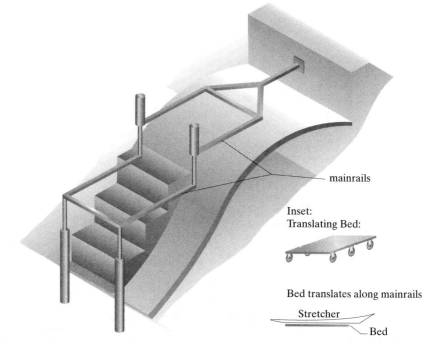

mainrails

Inset:
Translating Bed:

Bed translates along mainrails

Stretcher

Bed

Figure 8.9.

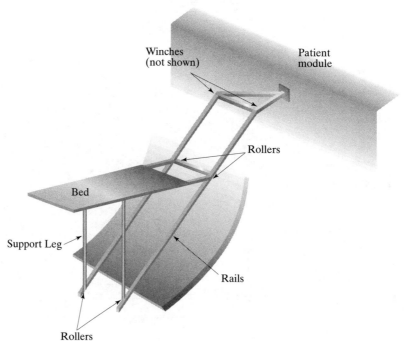

Winches
(not shown)

Patient
module

Rollers

Bed

Support Leg

Rails

Rollers

Figure 8.10.

The hydraulic lift design (design #3), shown in Figure 8.9, used hydraulic cylinders to raise and lower the stretcher. Two sets of main rails and a translating bed are employed by the design. This design overcame the disadvantages of the rocker arms in design #2, but introduced several new design flaws. These included misalignment of the rails at the top of the cylinder. Also, the position of the hydraulic cylinders interferes with the patient. Separation between the cylinders was calculated to be 16-inches, which would allow few patients to pass. Finally, the required stroke length of 20" would make necessary the use of heavy cylinders, violating design goal #4.

The fifth design employed twin rails, as illustrated in Figure 8.10, which make a 40° angle with the ground. Using rollers and a cable-and-winch system, the bed is moved from a lowered position to the opening of the aircraft. Vertical structural members support the bed as it moves and are also attached to the rails with rollers. This design met all the design goals and was selected as the final design.

After the final design was accepted, a kinematic and structural analysis was performed using 500 lb as the maximum weight of the patient. The goal was to move the patient quickly and efficiently with as little as possible discomfort. The analysis yielded production drawings and a bill of materials detailing the cost of each component in the design. This bill of material is reproduced in Table 8-1. An example of a detailed production drawing is shown in Figure 8.11.

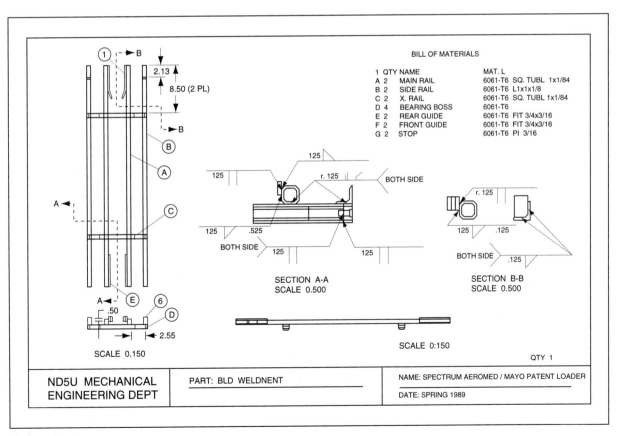

Figure 8.11.

TABLE 8-1   Bill of Materials and Cost Analysis

PART DESCRIPTION	QUANTITY	TOTAL COST (DOLLARS)
Ball screw assembly with bearing mount	1	348
Ball screw assembly three piece "hub" coupler	1	40
Motor	1	565
Dynetic 600 control (motor speed control)	1	391
Motor mounting plate	1	6
Motor bracket	1	1
Shim plate	1	1
Module arm bracket	2	2
Upper rail	2	28
Lower rail	2	28
Rail hinge	2	3
Bed rail	2	28
Load rod	1	2
Connecting arm	1	4
Vertical support leg	2	8
Support bracket	2	1
Support boss	2	1
Brace	1	2
Foot plate	1	2
Bed	1	17
Bed Boss	4	2
Roller bearing	8	100
Module attachment arm	1	19
Hardware		20
Labor		1000
Total Cost		$2619

8.5   SUMMARY

Engineering design is an iterative process. This design process will have ideation and refinement phases. In the ideation phase, the problem is defined; design goals detailed, and various solutions are examined. A preliminary design that meets the design goals is established. To communicate this design, rough sketches and models are produced.

In the refinement phase, the preliminary design is analyzed with respect to geometry, strength, and kinematic aspects. Detailed fully dimensioned drawings are produced for each component in the design.

Ruled paper, such as isometric paper, may be used to construct three-dimensional drawings of a model. CAD may be used to create detailed models and drawings. Parametric modelers, who define models in terms of parameters, are the preferred modelers, because they allow models to be easily updated.

KEY TERMS	CAD	engineering design
	concurrent engineering	traditional design

8.6 EXERCISES

E1. Discuss the differences between concurrent engineering and traditional design with respect to visual communications.

E2. Discuss the advantages of concurrent engineering with respect to a traditional-design approach.

E3. Discuss how CAD may be used in concurrent engineering.

E4. Define a parametic modeler.

E5. Define a constraint-based modeler.

E6. Discuss the differences between a feature-based modeler and a primitive modeler.

For the products in Problems E7–E13, list the design goals that a successful design would have to meet. What are some issues that may prevent the design from becoming economically, politically, or socially feasible?

E7. A hair dryer.

E8. An oscillating fan.

E9. A portable DVD player.

E10. A helmet for children soccer players.

E11. A new sports car.

E12. A manned spacecraft for a flight from earth orbit to Mars and return.

E13. A supersonic business jet.

Index